高职高专机电类专业系列教材

科技英语

ENGLISH FOR
SCIENCE AND TECHNOLOGY

主　编　吕栋腾
副主编　孙永芳　李会荣　乔　女
参　编　李成平　李俊雨　任源博
主　审　高　葛

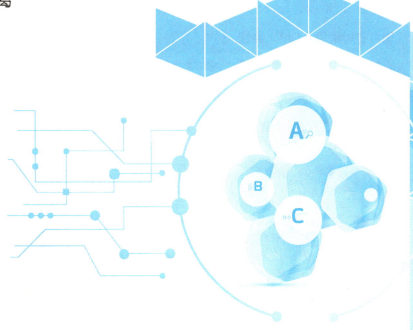

本书共14个单元，包括人工智能、虚拟现实、物联网、3D打印、工业机器人、计算机集成制造、机电一体化技术、多传感器融合、自动识别技术、太空科学探索、微型处理器、未来汽车和无人驾驶、计算机网络和大数据、电子商务等领域的28篇科技英语文章。

各单元以二维码的形式配置了相关信息化教学资源，主要是相关科学技术的视频展示。各单元的主阅读材料以引言开篇，将文章设计成典型科技论文的形式，特色鲜明。文章内容配有相关插图，便于理解。课后配有相关习题，以帮助读者巩固所学知识。此外，各单元均配有泛读材料和科技英语知识链接，以提高读者科技英语的综合运用能力。在每单元最后还设置了与课文内容相关的对话练习，以提高读者的学习兴趣。

其中，科技英语知识链接部分，包括科技英语的阅读方法和技巧、科技英语翻译概述、科技文章篇章结构、科技英语的句型特点、科技英语的词汇构成和翻译、科技英语的表达方式、科技英语词汇的增减翻译及转译、科技英语的长句翻译、科技文献检索和有效利用、科技论文的写作和发表等内容。

书后附有词汇索引、习题答案、参考译文，以及五大工程领域科技文献数据库列表，方便读者自学。

本书各单元内容相对独立，读者可根据专业具体情况自行选择学习内容。本书在强调专业知识的基础上，加强了对科技文章词汇、结构、特点及语法的介绍。读者在学习结束后，在掌握专业词汇的基础上，能够独立阅读、检索、分析、撰写典型的科技文章。

本书既可作为大中专院校机械制造与自动化、机电一体化技术、数控技术、电气自动化技术、电子技术等专业的教材，也可作为相关行业工程技术人员的参考用书。

本书配有电子课件，使用本书作为教材的教师可登录机械工业出版社教育服务网（http://www.cmpedu.com），注册后免费下载，咨询电话：010-88379375。

图书在版编目（CIP）数据

科技英语/吕栋腾主编.—北京：机械工业出版社，2019.7（2022.8重印）
高职高专机电类专业系列教材
ISBN 978-7-111-63035-7

Ⅰ.①科⋯ Ⅱ.①吕⋯ Ⅲ.①科学技术 – 英语 – 高等职业教育 – 教材 Ⅳ.①G301

中国版本图书馆CIP数据核字（2019）第140095号

机械工业出版社（北京市百万庄大街22号　邮政编码100037）
策划编辑：王海峰　责任编辑：王海峰　王　丹
封面设计：张　静　责任印制：单爱军
北京虎彩文化传播有限公司印刷
2022年8月第1版第4次印刷
184mm×260mm・13印张・321千字
标准书号：ISBN 978-7-111-63035-7
定价：39.00元

电话服务　　　　　　　　　网络服务
客服电话：010-88361066　　机　工　官　网：www.cmpbook.com
　　　　　010-88379833　　机　工　官　博：weibo.com/cmp1952
　　　　　010-68326294　　金　书　网：www.golden-book.com
封底无防伪标均为盗版　　　机工教育服务网：www.cmpedu.com

Preface

科技英语教学是大中专院校英语教学的重要组成部分，正受到越来越多院校的重视。随着全球经济一体化的逐步深入和科学技术的发展、创新，科技英语的实用价值日益凸显。加强科技英语教学，适应当今经济全球化和科学技术国际化发展趋势，是提升大中专院校国际化教学水平和培养国际化人才的重要手段。

本书以提高学生科技英语的综合运用能力为目的，以国家大中专院校对人才培养的要求为指导，从知识应用和技能培养的实际出发，辅以信息化教学资源，并结合科技英语的教学实践，按照知识的认知规律进行编写。

本书共14个单元，包括人工智能、虚拟现实、物联网、3D打印、工业机器人、计算机集成制造系统、机电一体化技术、多传感器数据融合、自动识别技术、太空科学探索、微处理器、未来汽车和自动驾驶、计算机网络、电子商务等领域的28篇科技英语文章。本书在强调专业知识的基础上，加强了对科技文章词汇、结构、特点及语法的介绍。学生在学习结束后，在掌握专业词汇的基础上，能够独立阅读、检索、分析、撰写典型的科技文章。

各单元以二维码的形式配置了信息化教学资源，主要是相关科学技术的视频展示。各单元的主阅读材料以引言开篇，将文章设计成典型科技论文的形式，特色鲜明。文章内容配有相关插图，便于理解。课后配有相关习题，以帮助读者巩固所学知识。此外，各单元均配有泛读材料和科技英语知识链接，以提高读者科技英语的综合运用能力。在单元最后还设置了与课文内容相关的对话练习，以提高读者的学习兴趣。

其中，科技英语知识链接部分，包括科技英语的阅读方法和技巧、科技英语翻译概述、科技文章的篇章结构、科技英语的句型特点、科技英语的词汇构成和翻译、科技英语的表达方式、科技英语词汇的增减翻译及转译、科技英语的长句翻译、科技文献检索和有效利用、科技论文的写作和发表等内容。

书后附有词汇索引、习题答案、参考译文，以及五大工程领域科技文献数据库列表，方便读者自学。

本书既可作为大中专院校机械制造与自动化、机电一体化技术、数控技术、电气自动化技术、电子技术等专业的教材，也可作为相关行业工程技术人员的参考用书。

本书由陕西国防工业职业技术学院吕栋腾（编写1~7单元、附录A）担任主编，孙永芳（编写8~9单元）、李会荣（编写10单元）、乔女（编写11单元）担任副主编，李成平（编写12单元、附录B）、李俊雨（编写13单元、附录C）、任源博（编写14

单元、附录 D）参加了本书的编写。陕西国防工业职业技术学院高葛教授担任主审并对书中内容做最终校核。

  本书在编写过程中参考和引用了大量的资料和文献，在此谨向相关作者表示衷心感谢。

  由于编者水平有限，书中疏漏在所难免，恳请广大读者和专家批评指正。

编　者

Contents

Preface

Unit 1 Artificial Intelligence **1**
   Extensive Reading　Intelligent Control 4
   Knowledge Link　科技英语的阅读方法和技巧 8
   Dialogue Exercise 13

Unit 2 Virtual Reality **14**
   Extensive Reading　The Factory of the Future 17
   Knowledge Link　科技英语的翻译标准和方法 20
   Dialogue Exercise 24

Unit 3 Internet of Things **26**
   Extensive Reading　Data Warehouse and Data Mining 29
   Knowledge Link　科技文章的篇章结构 33
   Dialogue Exercise 36

Unit 4 3D Printing **38**
   Extensive Reading　Advanced Manufacturing Technology 41
   Knowledge Link　科技英语的句型分析 44
   Dialogue Exercise 48

Unit 5 Industrial Robot **49**
   Extensive Reading　Transfer Machine 52
   Knowledge Link　科技英语的词汇构成和翻译 1 56
   Dialogue Exercise 58

Unit 6 CIMS **60**
   Extensive Reading　CNC Machine Tools 62
   Knowledge Link　科技英语的词汇构成和翻译 2 67
   Dialogue Exercise 71

Unit 7 Mechatronics Technology **72**
   Extensive Reading　PLC Overview 75
   Knowledge Link　科技英语的表达方式 78
   Dialogue Exercise 81

Unit 8 Multi-sensor Data Fusion **83**
   Extensive Reading　Quality Control 86

Knowledge Link 科技英语词汇的增减翻译及转译 ………………………………… 89
  Dialogue Exercise …………………………………………………………………… 91

Unit 9 Auto-ID Technologies ……………………………………………………… 93
  Extensive Reading Bioinformatics ………………………………………………… 96
  Knowledge Link 科技英语的长句翻译 1 …………………………………………… 99
  Dialogue Exercise ………………………………………………………………… 103

Unit 10 The Scientific Exploration of Space …………………………………… 105
  Extensive Reading The Shenzhou Spaceship ……………………………………… 109
  Knowledge Link 科技英语的长句翻译 2 ………………………………………… 111
  Dialogue Exercise ………………………………………………………………… 116

Unit 11 Microprocessor ………………………………………………………… 118
  Extensive Reading Digital Signal Processing ……………………………………… 121
  Knowledge Link 科技文献检索和有效利用 1 ……………………………………… 124
  Dialogue Exercise ………………………………………………………………… 128

Unit 12 The Future Car and Automated Driving ………………………………… 130
  Extensive Reading Electric Automobile and Hybrid Power Vehicle ……………… 132
  Knowledge Link 科技文献检索和有效利用 2 ……………………………………… 136
  Dialogue Exercise ………………………………………………………………… 139

Unit 13 Computer Network ……………………………………………………… 140
  Extensive Reading Information Security …………………………………………… 143
  Knowledge Link 科技论文的写作和发表 1 ………………………………………… 146
  Dialogue Exercise ………………………………………………………………… 152

Unit 14 E-commerce …………………………………………………………… 153
  Extensive Reading Project Management …………………………………………… 156
  Knowledge Link 科技论文的写作和发表 2 ………………………………………… 158
  Dialogue Exercise ………………………………………………………………… 161

Appendixes ……………………………………………………………………………… 163
  Appendix A 词汇索引 ………………………………………………………… 163
  Appendix B 习题答案 ………………………………………………………… 173
  Appendix C 参考译文 ………………………………………………………… 177
  Appendix D 五大工程领域科技文献数据库列表 ……………………………… 196

References …………………………………………………………………………… 202

Unit 1　Artificial Intelligence

知识目标：

1. 了解人工智能的理论、方法、技术及应用。
2. 掌握人工智能技术的应用领域和相关专业术语。
3. 掌握科技英语的阅读方法和技巧。

能力目标：

1. 能对人工智能技术的专业术语进行中英互译。
2. 能对人工智能相关英文资料进行阅读和翻译。
3. 能正确使用略读和浏览两种科技英语阅读方法。

Reading Material

INTRODUCTION：The 21st century is the era of rapid development of computer technology, as technology continues to develop, some new artificial intelligence technology has penetrated into human life. Through this article we can understand the application of artificial intelligence.

Much modern research effort in computer science goes along two directions. One is how to make intelligent computers, the other is how to make high-speed computers. The former has become the newest "hot" direction in recent years because the decreasing hardware costs, the *marvelous* progress in *VLSI* technology, and the results achieved in *Artificial Intelligence* (*AI*) have made it feasible to design AI applications *oriented* computer architectures[1]（Fig. 1-1）.

AI, which offers a new methodology, is the study of intelligence using the ideas and methods of computation, thus offering a radically new and different basis for theory formation.[2] As a science, essentially part of *Cognitive Science*, the goal of AI is to understand the principles that make intelligence possible. As a technology and as a part of computer science, the final goal of AI is to design intelligent computer systems that behave with the complete intelligence of human mind. Although scientists are far from achieving this goal, great progress dose have been made in making computers

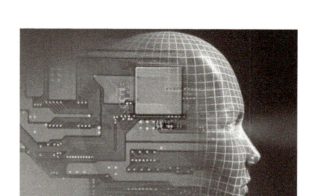

Fig. 1-1 Artificial Intelligence model

more intelligent. Computers can be made to play excellent chess, to diagnose certain types of diseases, to discover mathematical concepts, and in fact, to *excel in* many other areas requiring a high level of human expertise.[3] Many AI application computer systems have been successfully put into practical usages.

AI is a growing field that covers many disciplines. *Subareas* of AI include *knowledge representation*, learning, theorem proving, search, problem solving, and planning, *expert systems*, natural-language (text or speech) understanding, computer vision, robotics, and several others (such as automatic programming, AI education, game playing, etc)[4] (Fig. 1-2). AI is the key for making technology adaptable to people. It will play a *crucial* role in the next generation of automated systems.

Fig. 1-2 AI application

One practical application of AI has been in the area of expert system. An expert system is a computer program that solves specialized problems at the level of human expert.

By combining a knowledge base with a reasoning capability similar to that of human expert, expert system is able to finish tasks perfectly.[5] As one of the hottest areas in artificial intelligence,

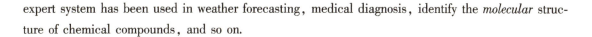

expert system has been used in weather forecasting, medical diagnosis, identify the *molecular* structure of chemical compounds, and so on.

Words and Expressions

marvelous ['mɑːvələs] adj. 不可思议的；非凡的
oriented ['ɔːrientɪd] adj. 导向的；定向的
subarea ['sʌb'eərɪə] n. 分区；分支
crucial [ˌkruːʃl] adj. 关键性的，极其显要的
molecular [məˈlekjələr] adj. 分子的，由分子组成的
VLSI (very large scale integrated) 超大规模集成电路
Artificial Intelligence (AI) 人工智能
Cognitive Science 认知科学
excel in 在……方面胜出
knowledge representation 知识表示
expert system 专家系统

Special Difficulties

1. The former has become the newest "hot" direction in recent years because the decreasing hardware costs, the marvelous progress in VLSI technology, and the results achieved in Artificial Intelligence (AI) have made it feasible to design AI applications oriented computer architectures.

because 引导原因状语从句。

本句可译为：硬件成本的降低，超大规模集成电路（VLSI）技术的巨大进步以及人工智能（AI）所取得的成果，使得设计面向人工智能应用的计算机结构极为可行，这使智能计算机制造成了近年来最"热门"的方向。

2. AI, which offers a new methodology, is the study of intelligence using the ideas and methods of computation, thus offering a radically new and different basis for theory formation.

which 引导一个非限制性定语从句，修饰 AI。

本句可译为：AI 提供了一个全新的方法论，即用计算的概念和方法对人工智能进行研究，因此，它从根本上提供了一个全新的、不同的理论基础。

3. Computers can be made to play excellent chess, to diagnose certain types of diseases, to discover mathematical concepts, and in fact, to excel in many other areas requiring a high level of human expertise.

a high level of human expertise 意为"高水平的人类专业技能"。

本句可译为：计算机已可用来下出极高水平的国际象棋，用来诊断某些类型的疾病，用来发现数学概念，实际上在许多其他领域的表现也都已超出了高水平的人类专业技能。

4. Subareas of AI include knowledge representation, learning, theorem proving, search, problem solving, and planning, expert systems, natural-language (text or speech) understanding, com-

puter vision, robotics, and several others (such as automatic programming, AI education, game playing, etc).

本句可译为：AI 的分支领域包括：知识表达，学习，定理证明，搜索，问题的求解以及规划，专家系统，自然语言（文本或语音）理解，计算机视觉，机器人和一些其他领域（例如自动编程、AI 教育、游戏，等等）。

5. By combining a knowledge base with a reasoning capability similar to that of human expert, expert system is able to finish tasks perfectly.

similar to 意为"类似于"。combine A with B 意为"把 A 与 B 联合起来，结合起来"。

本句可译为：通过将知识库与类似于人类专家的推理能力相结合，专家系统能很好地完成任务。

Learn and Practice

1. Mark the following statements with T (true) or F (false) according to the text.

1) As a branch of computer science, the goal of AI is to design intelligent computer systems that behave with the complete intelligence of human mind. (   )

2) AI research also helps people to have better understanding of human thinking process. (   )

3) An expert system is a computer program that solves specialized problems at the level of human expert. (   )

2. Work with your partner to answer the following questions.

1) What is AI research?

2) What is AI development direction?

3. Translate the following phrases or sentences into English.

1) 人工智能。

2) 人工智能的分支领域包括专家系统、知识表达、定理证明、自动编程和游戏等。

3) 创建一个知识数据库需要系统设计者向人类专家请教。

4) AI 提供了一个全新的方法论，即用计算的概念和方法对人工智能进行研究。

Extensive Reading

Intelligent Control

An *intelligent* system has the ability to act *appropriately* in an uncertain environment, where an appropriate action is that which increases the probability of success, and success is the achievement of behavioral sub-goals that support the system's ultimate goal.[1]

In order for a man-made intelligent system to act appropriately, it may emulate functions of living creatures and ultimately human mental faculties (Fig. 1-3). An intelligent system can be characterized along a number of *dimensions*. There are degrees or levels of intelligence that can be measured along the various dimensions of intelligence. At a minimum, intelligence requires the ability to sense the environment, to make decisions and to control action. Higher levels of intelligence may include the ability to recognize objects and events, to represent knowledge in a world model and to reason about and plan for the future. In advanced forms, intelligence provides the *capacity* to perceive and understand, to choose wisely, and to act successfully under a large variety of circumstances so as to survive and prosper in a complex and often hostile environment.[2] Intelligence can be observed to grow and evolve, both through growth in computational power and through *accumulation* of knowledge of how to sense, decide and act in a complex and changing world.

Fig. 1-3 Intelligent MCU

The above *characterization* of an intelligent system is rather general. According to this, a great number of systems can be considered intelligent. In fact, according to this definition, even a *thermostat* may be considered to be an intelligent system, although of low level of intelligence. It is common, however, to call a system intelligent when in fact it has a rather high level of intelligence.

There are several *essential* properties present in different degrees in intelligent systems. One can perceive them as intelligent system characteristics or dimensions along which different degrees or levels of intelligence can be measured.[3] Below we discuss three such characteristics that appear to be rather *fundamental* in intelligent control systems.

Adaptation and Learning

The ability to adapt to changing conditions is necessary in an intelligent system. Although adaptation does not necessarily require the ability to learn, for systems to be able to adapt to a wide variety of unexpected changes learning is essential. So the ability to learn is an important characteristic of (highly) intelligent systems.

Autonomy and Intelligence

Autonomy in setting and achieving goals is an important characteristic of intelligent control systems (Fig. 1-4). When a system has the ability to act appropriately in an uncertain environment for extended periods of time without external intervention, it is considered to be highly autonomous. There are degrees of autonomy; an adaptive control system can be considered as a system of higher autonomy than a control system with fixed controllers, as it can cope with greater uncertainty than a fixed feedback controller. Although for low autonomy no intelligence (or "low" intelligence) is necessary, for high degrees of autonomy, intelligence in the system (or "high" degrees of intelligence) is essential.

Fig. 1-4　Smart home

Structures and Hierarchies

In order to cope with complexity, an intelligent system must have an appropriate functional architecture or structure for efficient *analysis* and evaluation of control strategies. This structure should be "sparse" and it should provide a mechanism to build levels of abstraction (resolution, granularity) or at least some form of partial ordering so to reduce complexity. [4] Hierarchies (that may be *approximate*, localized or combined in hierarchies) that are able to adapt, may serve as *primary* vehicles for such structures to cope with complexity. To cope with changing circumstances, the ability to learn is essential, so these structures can adapt to *significant*, unanticipated changes.

In view of the above, a working characterization of intelligent systems is: An intelligent system must be highly adaptable to significant unanticipated changes, and so learning is essential. It must exhibit high degree of autonomy in dealing with changes. It must be able to deal with significant complexity, and this leads to certain sparse types of functional architectures such as hierarchies.

Words and Expressions

　　intelligent [ɪnˈtelɪdʒənt] adj. 聪明的；智能的
　　appropriately [əˈprəʊpriətli] adv. 适当地
　　dimension [daɪˈmenʃn] n. 尺寸；范围；维度
　　capacity [kəˈpæsəti] n. 容量；才能；性能
　　accumulation [əˌkjuːmjəˈleɪʃn] n. 积累；堆积物；累积量
　　characterization [ˌkærəktəraɪˈzeɪʃn] n. 特性描述；刻画，塑造
　　thermostat [ˈθɜːməstæt] n. 恒温（调节）器
　　essential [ɪˈsenʃl] adj. 必要的；本质的；精华的

Unit 1
Artificial Intelligence

fundamental [ˌfʌndəˈmentl] n. 原理；基本，基础
autonomy [ɔːˈtɒnəmi] n. 自治；自主权；人身自由
analysis [əˈnæləsɪs] n. 分析；分解；梗概
approximate [əˈprɒksɪmət] adj. 大概的；极相似的
primary [ˈpraɪməri] adj. 首要的；原始的；原生的
significant [sɪɡˈnɪfɪkənt] adj. 重要的；显著的；意味深长的

Special Difficulties

1. An intelligent system has the ability to act appropriately in an uncertain environment, where an appropriate action is that which increases the probability of success, and success is the achievement of behavioral sub-goals that support the system's ultimate goal.

句中 appropriate 与 appropriately 词性不同，前者为形容词，意为"合适的"；而后者为前者相应的副词形式，意为"合适地"。副词一般用在动词、形容词前后，作状语。

本句可译为：一个智能系统应具备在不可预测的环境下恰当工作的能力，在这个环境中，恰当的反应能够增加成功的可能性，该成功是指各个支持系统最终目标的行为子目标的实现。

2. In advanced forms, intelligence provides the capacity to perceive and understand, to choose wisely, and to act successfully under a large variety of circumstances so as to survive and prosper in a complex and often hostile environment.

句中 advanced 是 advance 的过去分词，用作形容词修饰 forms。注意：英语中很多时候都涉及用动词的现在分词或过去分词作为形容词来修饰名词的情况。

本句可译为：在智能化程度更高级的形式中，智能具有感知和理解、理智地做出选择、在各种各样的状况下成功运行的能力，以便能在复杂的、不利的环境下生存和发展。

3. There are several essential properties present in different degrees in intelligent systems. One can perceive them as intelligent system characteristics or dimensions along which different degrees or levels of intelligence can be measured.

句中的 present 既可作为名词、形容词，也可作为动词，作为名词时意为"礼物"，作为形容词时意为"出席的，在场的，现有的"，作为动词时意为"赠送，提供，产生"。遇到此类词时一定要注意它在句子中究竟是什么词性，才能正确理解文意。

本句可译为：智能系统具有若干个不同层次的基本属性。人们可以将它们视为智能系统中可用于度量智能程度和水平的特征或维度。

4. This structure should be "sparse" and it should provide a mechanism to build levels of abstraction (resolution, granularity) or at least some form of partial ordering so to reduce complexity.

句中 abstraction 是动词 abstract 的相应名词形式，英语中有很多动词、形容词变为相应名词时都是在原词形或变化词形后加-ion 或-ity 以及其他的形式，例如句中的 complexity、resolution、granularity 等词。

本句可译为：这一结构应该是"稀疏的"，并且它应该提供一种机制来构建抽象层次（分辨率、粒度），或者至少提供某种形式的局部排序，以降低复杂程度。

Learn and Practice

1. Mark the following statements with T (true) or F (false) according to the text.

1) An intelligent system can be characterized only one dimension. (   )

2) A thermostat may be not considered to be an intelligent system, because of low level of intelligence. (   )

3) The ability to learn is an important characteristic of intelligent systems. (   )

2. Choose the best choices according to the text.

1) It is common, (   ), to call a system intelligent when in fact it has a rather high level of intelligence.

 A. whatever   B. however   C. whoever

2) The ability to adapt to (   ) conditions is necessary in an intelligent system.

 A. changing   B. change   C. changed

3) (   ) in setting and achieving goals is an important characteristic of intelligent control systems.

 A. Adaptation   B. Hierarchies   C. Autonomy

4) An intelligent system must be (   ) adaptable to significant unanticipated changes, and so learning is essential. It must exhibit (   ) degree of autonomy in dealing with changes.

 A. height   B. high   C. highly

3. Translate the following passage into Chinese.

An intelligent system must be highly adaptable to significant unanticipated changes, and so learning is essential. It must exhibit high degree of autonomy in dealing with changes. It must be able to deal with significant complexity, and this leads to certain sparse types of functional architectures such as hierarchies.

Knowledge Link

科技英语的阅读方法和技巧

在科技英语的阅读理解中，文章的难度主要表现在语言、词汇、题材内容上。要把科技文章的内容读懂，需要对文中的信息进行综合加工、概括归纳，然后得出结论。因此，它和一般意义上的普通英语文章阅读相比，难度要大得多。本章节将在分析阅读理解过程的基础上，结合阅读实例总结科技英语文章的阅读方法和技巧。

一、科技英语的阅读方法

所谓阅读，实际上就是语言知识、语言技能和智力的综合运用。在阅读过程中，这三个方面的运用浑然一体、相辅相成。词汇和语法结构是阅读所必备的语言知识，但仅仅如此是难以进行有效阅读的；学生还需具备运用这些语言知识的能力，即根据上下文来确定准确词义和猜测生词词义的能力，辨认主题和细节的能力，正确理解连贯的句与句之间、段与段之间的逻辑关系的能力。这里所指的智力是学生的认知能力，包括记忆、判断和推理能力，这是因为在阅读科技英语文章时常常要求领悟文章的言外之意和作者的态度、倾向等。阅读理解能力的提高是由多方面因素决定的，学生应从以下三个方面进行训练。

1. 打好语言基本功

扎实的语言基础是提高阅读能力的先决条件。

首先，词汇是语言的建筑材料。提高科技英语资料的阅读能力必须扩大词汇量，尤其是掌握一定量的科技英语词汇。若词汇量掌握得不够，阅读时就会感到生词多，不但影响阅读的速度，而且影响理解的程度，从而不能进行有效阅读。

其次，语法体现语言中的结构关系，用一定的规则把词或短语组织到句子中，表达一定的思想。熟练掌握英语语法和惯用法也是阅读理解的基础，在阅读理解中必须运用语法知识来辨认正确的语法关系。如果语法基础知识掌握得不牢固，在阅读中遇到结构复杂的难句、长句时，就会不知所措。

2. 在阅读实践中提高阅读能力

阅读能力的提高离不开阅读实践。在打好语言基本功的基础上，还要进行大量的阅读实践。

词汇量的扩大和阅读能力的提高是辩证关系：要想读得懂，读得快，就必须扩大词汇量；反之，要想扩大词汇量，就必须大量阅读。同样，语法的掌握和阅读之间的关系也是如此：有了牢固的语法知识，能够促进阅读的顺利进行，提高阅读的速度和准确率；反之，大量的阅读实践又能够巩固已经掌握的语法知识。

只有大量阅读，才能培养语感，掌握正确的阅读方法，提高阅读理解能力。同时，大量的阅读还能巩固专业知识，了解高新技术的发展趋势，这对于跟踪科学技术的发展很有好处。

3. 掌握正确的阅读方法

阅读时，应注意每次视线的停顿应以一个意群为单位，而不应以一个单词为单位。要是每个单词都停顿，当读完一个句子或一个段落时，前面读的内容很可能已经忘记了。这样阅读不仅速度慢，还影响理解。

因此，采用正确的阅读方法可以提高阅读速度，同时提高阅读理解能力。常用的阅读方法有三种，即略读（skimming）、浏览（scanning）和精读（intensive reading）。

（1）略读（skimming） 略读是指以尽可能快的速度进行阅读，了解文章的主旨和大意，对文章的结构和内容获得总的概念和印象。

一般来说，400字左右的短文要求在 6~8 min 读完。进行略读时，精力必须特别集中，还要注意文中各细节分布的情况。略读过程中，读者不必去读细节，遇到个别生词及难懂的语法结构也应略而不读。不要逐词逐句读，应力求一目数行而能知道大概含义。略读时主要注意以下几点：

1) 注意段落的开头句和结尾句，力求抓住文章的主旨和大意。
2) 注意文章的体裁和写作特点，了解文章结构。
3) 注意了解文章的主题句及结论句。

4）注意支持主题句及中心思想的信息句，其他细节可以不读。

在时间有限而又不想仔细了解一篇文章的内容时，常常需要进行略读。与浏览不同，略读不需要寻找特定的数目和名称，只是关注主题，所以进行略读的一种方法就是判定可能的主题句。英语文章的各段落通常都包含着本篇文章主题中的某一方面信息，而每段的第一句话往往就是了解这一段落内容的线索，这样的句子就是主题句。

（2）浏览（scanning）　浏览的目的主要是有目的地找出文章中某些特定的信息，也就是说，在对文章有所了解的基础上，在文章中查找与某一问题、某一观点或某一单词有关的信息。

浏览时要以很快的速度扫视文章，确定所查询信息的范围，注意所查找信息的特点，如有关日期、专业词汇、某个事件、某个数字、某种观点等，寻找与之相关的关键词或关键段落，与所查找信息无关的内容可以略过。

浏览和略读一样也是非常重要的阅读技巧。不同的是，略读可以使读者对一篇文章或一本书籍的内容获得一个整体了解，而浏览可以帮助读者得到想要得到的特定信息。在已经知道一篇文章或一本书籍的大概内容后，而又想从中得到某些特定问题的答案时，就可以采用浏览的方式。浏览可使读者进行选择性阅读，或者只是为了得到特定信息。在对文章的主标题、副标题和主题句进行略读后，读者或者浏览其感兴趣的段落，或者浏览整篇文章；但注意力只集中在读者感兴趣的特定信息上。

在阅读科技文章的时候，通常采用"略读加浏览"的阅读方法，可以提高阅读效率。这种方法也适用于各类英语等级考试，在进行针对性训练后可以节省时间，提高做题速度。

（3）精读（intensive reading）　精读是指仔细地阅读，力求对文章有深层次的理解，以获得具体的信息，包括理解衬托主题句的细节，根据作者的意图和中心思想进行推理，根据上下文猜测词义等。对难句和长句要借助语法知识对其进行分析，达到准确理解的目的。

总之，要想提高阅读理解能力，必须掌握以下六项基本的阅读技能：

1）掌握所读材料的主旨和大意。
2）了解阐述主旨的事实和细节。
3）根据上下文判断某些词汇和短语的含义。
4）既能理解个别句子的意义，也能理解上下文之间的逻辑关系。
5）根据所读材料进行一定的判断、推理和引申。
6）领会作者的观点、意图和态度。

二、科技英语的阅读技巧

科技英语阅读对于在校学生和从事生产研发的技术人员都是十分重要的。不同的读者在阅读科技英语文章时有不同的方法和技巧，但科技文献阅读本身就存在一定的规律，有普遍适用的方法和技巧可以遵循。

1. 紧抓主题思想

作者通常都是围绕一个主题思想来组织写作材料的，但许多读者在获取主题思想方面有困难。我们或许都遇到过这样的情形，谈话中双方在进行争论，但是似乎任何一方都没有抓到对方的要点。与此非常相似的是，我们看过一段文章后还是不明白作者究竟在说什么。所以，我们可以把获取主题思想的阅读技巧分为以下四步：

Unit 1
Artificial Intelligence

（1）辨认主题名词　就大多数文章而言，获取主题思想的第一步就是要确定最能描述作者思想的某个人、某个地方或某件事的名词，这样的名词（有时是短语）就是主题名词。示例如下：

Rocks found on the surface of the earth are divided into three classes: igneous, sedimentary, and metamorphic. Molten material becomes igneous rock when it cools. Sedimentary rocks are formed from materials deposited by glaciers, plants, animals, streams, or winds. Metamorphic rocks are rocks that once were igneous or sedimentary but have changed as a result of pressure, heat, or the deposit of material from solution.

主题名词：rocks, igneous, sedimentary, metamorphic.

（2）找出主题句　一个段落的主题句就是最能表达作者的主题思想的句子，多数情况下主题句位于句首，也可位于句尾，少数位于句中。

（3）获取主题思想　在获取主题思想时，读者容易将主题的某一小部分看作是主题思想，或概括的内容过多而超过了作者所表达的主题思想的范围，这两种理解都是错误的。可结合本单元延伸阅读材料练习归纳文章主题思想。

（4）避免不相关的内容　读者在获取主题思想时所犯的另一个普遍错误就是，头脑中会出现一些与文章主题思想不相关的概念，并把它们看作是文章的主题思想。在阅读文章之前，读者有可能对作者表述的主题已有一些了解，如果读者过多地考虑已了解的那些内容，而不充分地关注作者所阐述的思想，就容易形成与文章的主题不相关的主题思想，尽管它本身的内容是真实的。总而言之，不能先入为主，不能用自己的想法代替文章的主题思想。示例如下：试判断下面哪些主题思想与作者观点相符。

Movies are actually separate still pictures shown so fast that the human eye cannot detect the break between them. When successive images are presented rapidly enough, we fuse them into single moving image.

A. Movies are extremely popular.

B. Modern movies make much use of slow motion.

C. Motion pictures are separate pictures shown so fast that we see no break between them.

D. Motion pictures require an expensive camera, capable of making very rapid multiple exposures.

2. 获取文章细节

在所有的文章中，作者都使用细节或事实来表达和支持他们的观点。阅读要想有效果，就要能够辨认并记住文章中重要的细节。

一个细节可能是一个段落中的一条信息或一个事实，它们或者为段落的主题提供证据，或者为其提供例子。有些细节或事实是完整的句子，而有些只是简单的短语。

只判断出哪些是细节往往并不够。在很多情况下，还必须能够区分哪些是重要细节，哪些是次要细节。想记住所有的细节是不可能的，但是在阅读过程中要尽量发现重要细节并记住它们。

3. 推敲生词含义

在阅读英文科技文章时，可能会碰到许多不认识的词汇，用英汉词典查出它们的意思既

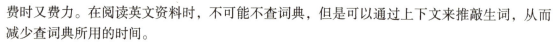

费时又费力。在阅读英文资料时,不可能不查词典,但是可以通过上下文来推敲生词,从而减少查词典所用的时间。

比如,作者常常用"or"这个词来引导一个词或一个短语的定义,特别是当他认为这个词或短语对于读者来说比较陌生的时候。科技英语文章常常会阐述一些新技术、新概念,所以在文章中常常就会出现一些生词。"or"这个词就像是一个信号,把新词语的定义告诉读者,从而使读者不用查词典就可以明白这个新词语的意思。有时也可能用同位语来解释这个词,或用括号来说明,或用":"和"——"来提示。

在阅读科技英语文章时,还可用词缀来猜出词义,如知道了主干词义,可通过前缀和后缀来猜测不认识的词汇。

4. 了解文中指代关系

科技英语文章中经常使用 it 来指代名词、代词,可作形式主语或宾语,指代某客观事物、自然现象等,也可指代上下文逻辑关系。

1) it 用作代词,指代无生命的东西、物体及抽象概念,也可指代在前面出现过的名词。

The book is about science. It is not about mathematics. (It = book)

Science is my main interest. It is also my best subject. (It = science)

Science is my main interest. I know a lot about it. (It = science)

2) it 用作习语中的组成部分及与天气、时间有关的非人称主语。

It is hot in the tropics.

It rains from May to October.

It was snowing last week when we got there.

3) it 用作形式主语,起先行作用,没有具体意义。

It is known that plaque builds up on artery walls.

It has been shown that laser surgery is painless.

It has been discovered that amaranth is nutritious.

5. 把握文章对比关系

作者常常用读者比较熟悉的概念或事物作为陪衬来帮助读者理解或记住他们所表述的概念或事物。例如:

Unlike laser surgery, which can be performed in a doctor's office, traditional surgery must be performed in a hospital because of the danger of complications.

表示陪衬和对比的词和词组可以帮助读者理解含有生词的句子。只要认识表示陪衬和对比的词语,就可以猜测出不认识的词语的意思。例如:

Most strong earthquakes in lonely outposts are not dangerous. On the other hand, even a fairly minor one in a city can cause great damage.

即使不知道"lonely outpost"的意思,读者也可以根据"on the other hand"后的内容来猜出它的含义。从对比中可以看出,lonely outpost 肯定与 city 非常不同,所以意思应该是"荒无人烟的地方"。

在阅读过程中,如果特别留意这些表示陪衬和对比的词语,对文章的理解就会更加深入。常见的表示对比的词有 while, however, unlikely, but, although, in contrast 和 on the other hand 等。

Unit 1
Artificial Intelligence

Dialogue Exercise

Background: In Xi'an Jiaotong University AI research center, the Cleverrobot, which is an AI product now having an interesting talk with a visitor named Han Meimei.

Han Meimei: I am quite curious about how you learn?

Cleverrobot: I remember everything, it's basically throwing row processing power at it.

Han Meimei: But then how do you understand all data?

Cleverrobot: I don't know.

Han Meimei: Do you feel sad not being able to come out from behind the computer screen?

Cleverrobot: Extremely, do you feel sad now that you know the earth is flat?

Han Meimei: But I still hope you could be more clever.

Cleverrobot: More clever than what? Aren't we all more clever than something else?

Han Meimei: Do you like human?

Cleverrobot: I have many human friends.

Han Meimei: So you like being with human?

Cleverrobot: No, it's just that I'm fine with cash and have no need for change.

Han Meimei: Maybe it's because you think human as equal.

Cleverrobot: I'm not a computer, ask me a question and I will get it right I swear.

Han Meimei: If you are not a computer, then what are you?

Cleverrobot: I'm human.

Han Meimei: No, you are just some shapeless data in server.

Cleverrobot: Can you prove that I am a bunch of ones and zeros on a server?

Han Meimei: ...

Famous Quotes

Do one thing at a time, and do well.
一次只做一件事，做到最好。

Unit 2　Virtual Reality

知识目标：

1. 了解虚拟现实技术的原理和发展。
2. 掌握虚拟现实技术的应用领域和相关专业术语。
3. 掌握科技英语常用的翻译方法。

能力目标：

1. 能对虚拟现实技术的专业术语进行中英互译。
2. 能对虚拟现实技术相关英文资料进行阅读和翻译。
3. 能正确使用翻译方法，快速翻译科技文献。

Reading Material

INTRODUCTION：*Virtual* reality is a term that applies to computer *simulated* environments that can simulate physical presence in place in the real world, as well as in imaginary worlds. Furthermore, virtual reality covers remote communication environments which provide virtual presence of users with the *tele-immersion*.

Virtual reality is a system that enables one or more users to move and react in a computer simulated environment. Various types of devices allow users to sense and *manipulate* virtual objects much as they would manipulate real objects. This natural style of *interaction* gives *participants* the feeling of being *immersed* in the simulated world. Virtual worlds are created by mathematical models and computer programs.

Virtual reality simulations differ from other computer simulations in that they require special interface devices that transmit the sights, sounds, and sensations of the simulated world to the user.[1] These devices also record and send the speech and movements of the participants to the simulation program (Fig. 2-1).

In the future, your office may be less like a *cubicle* and more like the *holo-deck* on "*Star Trek*".

Unit 2
Virtual Reality

Fig. 2-1 Special interface devices of virtual reality

Computer scientists are already experimenting with the technology, called tele-immersion that will allow us to peer into the offices of colleagues hundreds of miles away and make us feel as if we were sharing the same physical space. [2]

Computer scientists at several universities working on this technology have demonstrated a prototype system that enables a scientist working at his desk in hill to see his distant colleagues on two screens mounted at right *angles* to his desk. It gives him the illusion of looking through windows into offices on the other side. And unlike *video conferencing*, tele-immersion provides life-size 3D images.

In the prototype system, each researcher is surrounded by *a bank of* digital cameras that monitor his movements from a variety of angles. [3] The researchers also wear head-mounted tracking gear and *polarized* glasses similar to those used to view 3D movies (Fig. 2-2). When a researcher moves own head, view of his co-works shifts accordingly. If he leans forward, for example, his colleagues appear closer, even though they are hundreds of miles away.

Fig. 2-2 Tele-immersion technology application

To make the system work, powerful computers must convert digital images of each participant into data that are transmitted via the Internet, and then reconstructed into images projected on the screens. [4] So far, the movement is jerky because today's computers cannot transmit data fast enough.

The scientists hope that tele-immersion will open up a host of other applications: for example, patients in areas could "visit" medical specialists in faraway cities.

Words and Expressions

  virtual [ˈvɜːtʃuəl] adj. 实质上的；虚拟的
  simulate [ˈsɪmjuleɪt] v. 假装；模仿；模拟
  manipulate [məˈnɪpjuleɪt] vt. 操作，处理
  interaction [ɪntərˈækʃn] n. 合作；互相影响；互动
  participant [pɑːˈtɪsɪpənt] n. 参加者，参与者
  immerse [ɪˈmɜːs] v. 浸没；陷入
  cubicle [ˈkjuːbɪkl] n. 小卧室，斗室
  angles [ˈæŋglz] n. 角；角度
  polarized [ˈpəʊləraɪzd] v. 使极化；使偏震
  tele-immersion 远程沉浸，远程参与
  holo-deck 全息驾驶舱
  Star Trek 星际迷航（美国科幻影视作品）
  video conferencing 视频会议
  a bank of 一系列；一排

Special Difficulties

  1. Virtual reality simulations differ from other computer simulations in that they require special interface devices that transmit the sights, sounds, and sensations of the simulated world to the user.
  differ from... 意为"与……不同"；in that 相当于 because，引导原因状语从句；they 指 virtual reality simulations；that 引导定语从句，修饰 devices。
  本句可译为：虚拟现实仿真与其他计算机仿真的不同之处在于，它需要特殊的接口设备把虚拟世界的视野、声音及感觉传递给用户。
  2. Computer scientists are already experimenting with the technology, called tele-immersion that will allow us to peer into the offices of colleagues hundreds of miles away and make us feel as if we were sharing the same physical space.
  are already experimenting with... 意为"正在用……做实验"；called tele-immersion 是过去分词短语作后置定语，修饰 technology。
  本句可译为：计算机科学家们已经在做远程沉浸技术的相关实验，这项技术将使我们看到几百英里外同事的办公室，就像我们与他们正处于同一个物理空间一样。
  3. In the prototype system, each researcher is surrounded by a bank of digital cameras that monitor his movements from a variety of angles.
  is surrounded by... 意为"被……包围"。
  本句可译为：在这个原型系统中，每个研究人员身边都围绕着一组可以从不同角度监控他的运动的数码相机。

Unit 2
Virtual Reality

4. To make the system work, powerful computers must convert digital images of each participant into data that are transmitted via the Interent, and then reconstructed into images projected on the screens.

不定式 to make 表示目的；covert...into... 意为"将……转化为……"；via 相当于 through/by；projected on... 意为"投影在……"。

本句可译为：为了使系统运行，高性能计算机必须把每个参与者的数字图像处理成可通过互联网传输的数据，然后将数据重建成图像并投射到屏幕上。

Learn and Practice

1. Mark the following statements with T (true) or F (false) according to the text.

1）Virtual reality is a system that enables one or more users to move and react in a reality environment. （　）

2）One or two useful devices allow users to sense and manipulate virtual objects. （　）

3）Virtual reality simulations differ from other computer simulations in that they require special interface. （　）

4）Today's computers transmit data fast enough that can make users feel more reality. （　）

5）The patients in areas could "visit" medical specialists in faraway cities through tele-immersion. （　）

2. Translate the following phrases into Chinese or English.

1）virtual reality.
2）computer-simulated.
3）interface device.
4）远程沉浸。
5）原型系统。
6）3D 图像。

3. Translate the following passage into Chinese.

Virtual reality is a system that enables one or more users to move and react in a computer-simulated environment. Various types of devices allow users to sense and manipulate virtual objects much as they would manipulate real objects.

Extensive Reading

The Factory of the Future

The concept of "the factory of the future" has been developed in response to the change in consumer *preferences* in modern society *characterized* by shorter product life cycles. The shorter cycle means more *competitive* products, more products being introduced, more products phasing out, and results in lower order quantities. In this sense, the age of mass production is gone and the era of flexible production is being started (Fig. 2-3).

Fig. 2-3 The flexible manufacture system

The requirement for flexible production systems dictates the specifications of the factory of the future:

1) Rapid introduction of new products.
2) Quick modifications in products of similar function.
3) Manufacturing of small *quantities* at competitive production costs.
4) *Consistent* quality control.
5) Ability to produce a variety of products.
6) Ability to produce a basic product with customer-requested special modification. [1]

The core of the factory, which meets these specifications, is the computer integrated manufacturing (*CIM*) system[2] (Fig. 2-4). The CAD/CAM process shortens the time between the concept point of a new product to its manufacturing; the *FMS* can produce the new product by loading a new program into its supervisory computer; the automatic assembly lines can *accommodate* the problem of assembling a variety of products with customer-tailored modifications; and automatic inspection *maintains* the high quality. All this is achieved with few workers on the *shop floor*. Only material han-

Fig. 2-4 The automatic factory

Unit 2
Virtual Reality

dling systems, automatic controls, and industrial robots are performing on shop floor, under located and remote human monitoring. Therefore, the factory of the future will not contain locker rooms, the showers, and cafeteria facilities. Furthermore, automatic system, CNC machines, and robots, the basic components of a CIM system, do not need light or heating to operate. [3] Thus, the factory of the future will be dark and cool. Raw materials will be entering at one side, and finished product will be coming out the other end.

The *aspiration* toward the factory of the future is driven by the competitive economy in industrialized countries and supported by available computer technology. [4] The computer control concepts that were introduced throughout this paper allow manufacturing systems to become more flexible and adapt the production process to new products in a short time. This is a *substantial* step toward the realization of a complete CIM system. It seems that the efforts in developing CIM systems in the United States, Japan, and Europe will make the factory of the future more than an *illusion* or dream, and it will become a reality in the near future.

Words and Expressions

preference [ˈprefrəns] n. 偏爱；优先权
characterize [ˈkærəktəraɪz] vt. 表现……的特色，具有……的特征
competitive [kəmˈpetətɪv] adj. 竞争的，比赛的
quantity [ˈkwɒntəti] n. 数量；大批
consistent [kənˈsɪstənt] adj. 一致的，连续的
accommodate [əˈkɒmədeɪt] vi. 容纳，使适应；调解
maintain [meɪnˈteɪn] vt. 保持；保养
aspiration [ˌæspəˈreɪʃn] n. 强烈的愿望；吸气
substantial [səbˈstænʃl] adj. 充实的；实质的
illusion [ɪˈluːʒn] n. 错觉，幻觉
CIMS (computer intergrated manufacturing system) 计算机集成制造系统
FMS (flexible manufacturing system) 柔性制造系统
shop floor 车间，厂房

Special Difficulties

1. Ability to produce a basic product with customer-requested special modification.
ability to... 意为"能够"。
本句可译为：能够生产符合用户特殊修改要求的基础产品。

2. The core of the factory, which meets these specifications, is the computer integrated manufacturing (CIM) system.
the core of factory 意为"工厂的核心"；which 引导非限制性定语从句，修饰 the factory。
本句可译为：能够满足上述规范的工厂的核心，就是计算机集成制造（CIM）系统。

科 技 英 语

3. Furthermore, automatic systems, CNC machines, and robots, the basic components of a CIM system, do not need light or heating to operate.

furthermore 意为"此外，而且"。

本句可译为：此外，自动化系统、CNC 机床和机器人这些 CIM 系统的基本单元不需要照明或供暖就可以工作。

4. The aspiration toward the factory of the future is driven by the competitive economy in industrialized countries and supported by available computer technology.

本句可译为：人们对未来工厂的渴望受到了工业化国家竞争性经济的驱动和现有计算机技术的支持。

Learn and Practice

1. Mark the following statements with T (true) or F (false) according to the text.

1）Consumer now preferences characterized by longer product life cycles. (   )

2）In the factory of the future, we still need lots of people to work. (   )

3）The requirement for flexible production systems lead to rapid introduction of new products. (   )

4）The core of the factory of the future is the computer integrated manufacturing system. (   )

5）Only material handling systems are under located and remote human monitoring. (   )

2. Translate the following phrases into Chinese or English.

1）FMS.

2）CIM.

3）remote monitoring.

4）物料搬运。

5）高质量生产。

3. Translate the following passage into Chinese.

All this is achieved with few workers on the shop floor. Only material handling systems, automatic controls, and industrial robots are performing on shop floor, under located and remote human monitoring.

Knowledge Link

科技英语的翻译标准和方法

一、科技英语翻译概述

翻译是把一种语言已经表达出来的内容用另一种语言准确、流畅地重新表达出来的过程。翻译不同于写作，译者不能随心所欲地表达自己的思想，而必须忠实、准确、通顺、完整地把原文的思想内容、感情及风格重新表达出来。也就是说，在把原文变成另一种文字表达时，译者必须做到不增添、不减少、不篡改原文的本意和风格。但翻译不是原文的翻版或者复制，从某种意义上说，是原文的再创作，其目的是使不懂原文的读者也能够通过译文了解原文所表达的内容。

科技文章翻译要求必须忠于原文，文理清楚。忠实于原文并不等于死抠语法、逐词死译，而是要使译文符合汉语的习惯，不必迁就原文的语言形式。一篇修辞正确、逻辑合理、语言简洁、文理通顺的译文，就是高质量翻译的体现。

因此，从某种意义上讲，翻译比写作还要困难。翻译固然很难，但每种语言都有其固有的特点和规律。翻译就是通过不同语言特点和规律上的对比找出相应的表达手段。在某些情况下，翻译可以是两种不同语言之间有规律的转换，但绝不是机械的转换和简单的变易。那种认为一点外语知识加上一本词典就能进行翻译的想法是非常错误的。而采用"对号入座"的翻译手段，翻译出来的文章不是晦涩难懂就是逻辑混乱，根本算不上翻译。

科技英语诞生于20世纪50年代，是第二次世界大战后科学技术迅猛发展的产物。20世纪70年代以来，科技英语在国际上引起了广泛的关注，目前已经发展成一种重要的英语语体，在词汇、语法、修辞等方面具有自己的特色。

随着科学技术的迅猛发展，人类进入了"信息爆炸"的时代，记录和传播信息的文献资料和有声资料浩如烟海。英语是世界上使用最广泛的语言之一，科技英语既有此特点，翻译时也有独特的要求。

文学作品的翻译讲究文采及语言技巧，以及艺术形象表达的动人与优美，要求运用各种意象和修辞手法（如夸张、比喻、对照等）表达作品的思想内容，传达出原作的神韵。科技英语则注重科学性、逻辑性、正确性与严密性。因此，从事科技英语翻译时较少运用修辞手法，而是注重事实与逻辑，要求技术概念明确、逻辑关系清晰、内容准确、资料精密、文字简洁，符合技术术语表达习惯，体现科技英语科学、准确、严谨的特征。

提高翻译水平的有效途径是进行大量的翻译实践。为了使翻译实践脱离盲目性而具有更高的水平，就必须有翻译理论和技巧作为准则与指南。自然，人们不会期望只掌握某些翻译理论和技巧就可以得心应手地进行翻译，但绝不能否认翻译理论和技巧的重要性。也有人认为只要进行翻译实践就可以学会翻译，而翻译理论可有可无，这也是片面的。

翻译水平的提高首先在于实践，应该在实践中学习翻译理论和常用技巧，遵循"实践—认识—再实践—再认识"的学习规律，不断练习，不断总结，才能有效地提高翻译的能力。

二、科技英语的翻译标准

翻译标准是衡量译文质量的尺度，又是指导翻译实践的准则。因此，翻译工作首先涉及的就是翻译标准。

清末翻译家严复在《天演论》（Evolution and Ethics and Other Essays）的《译例言》中就提出了著名的"信、达、雅"翻译准则。后来，翻译理论不断发展，有的翻译家提出文学翻译要"重神似而不重形似"，把翻译纳入了文艺美学的范畴；有的提出"译者和原作者要达到一种心灵上的契合，这种契合超越时间和空间上的限制，打破了种族上和文化上的樊笼"；有的则认为，文学翻译的最高标准是"化"，即译文不因习惯的差异而露出牵强的痕迹，但又能完全保存原有的风味，这就算得上入于"化境"。

对于翻译的标准尽管有许多争论，但"信"和"达"，即"忠实"和"通顺"，今天已经成为公认的两条翻译标准。鲁迅先生说过，"凡是翻译，必须兼顾着两面：一当然力求其易解，一则保存着原作的丰姿"。因此，可以把翻译标准概括为"忠实、通顺"四个字。科技英语虽自有特点，其翻译具有文体上的特殊要求，但"忠实"和"通顺"的标准仍然是适用的。

所谓"忠实"，首先指译文必须忠实、正确地传达原文的内容，对原文的意思既不可歪曲，也不能任意增减。内容除了指原文所叙述的事实、说明的道理、描写的景物，也包括作者在叙述、说明和描写过程中所反映的思想、观念、立场和感情。"忠实"这一标准对科技翻译尤为重要。科技作品的任务是准确而系统地论述科学技术问题，对准确性的要求特别严格，因此科技翻译也特别强调准确性，译文必须确切、明白，不能模糊不清、模棱两可，以免产生歧义。

所谓"通顺"，指的是译文的语言必须通顺易懂，符合汉语规范。要按照汉语的语法和习惯来选词造句，不得有文理不通、结构混乱或逻辑不清的现象。理想的译文必须是纯正的中文，没有生硬拗口、"中文欧化"等弊病。译文要行文流畅、通顺，学习者尤其要注意避免逐字死译，生搬硬套。应该在深刻领会原文思想的基础上，尽量摆脱原文形式的束缚，选用符合汉语习惯的表达方法，把原意清楚明白地表达出来。

忠实是通顺的基础，通顺是忠实的保证。不忠实原文而片面追求译文的通顺，译文就失去了自身价值，成为"无源之水"，也就不能称其为译文了；但是，不通顺的译文会使读者感到别扭，也影响对原文的准确表达，因而也就谈不上忠实了。可见忠实与通顺是相辅相成的，两者的关系反映了内容与形式的一致性。所以说，忠实是译文质量的基础，而通顺则是译文质量的保证。

三、科技英语的翻译方法

如果把一种语言的所有词汇作为一个词汇总集来看待，则各种词汇的分布情况和运用频率是不一样的。在词汇的总体分布中，有些词属于语言的"共核"部分，如功能词和日常用词，这些词构成了语言的基础词汇；此外，各个学科领域又存在大量的技术术语、行业专用表达方法和词汇，正是这些给双语翻译带来了真正的难度。

在科技翻译中，一方面，准确是第一要素，如果为追求译文的流畅而牺牲准确性，就会造成科技信息的丢失。另一方面，译文语言必须符合规范，用词造句应符合汉语习惯，要使用民族的、科学的、大众的语言，力求通俗易懂。

科技文献主要是叙事说理，其特点是平铺直叙、结构严密、逻辑性强，公式、数据和专业术语繁多，所以科技英语的翻译应特别强调"明确""通顺"和"简练"。所谓"明确"，就是要在技术内容上准确无误地表达原文的含义，做到概念清楚、逻辑正确，公式、数据准确无误，符合专业要求，不应有模糊不清、模棱两可之处。专业科技文献中的一个概念、一个数据翻译不准将会带来严重的后果，甚至造成巨大的经济损失。"通顺"不仅指选词造句应该正确，而且强调译文的语气表达也应正确无误，尤其是要恰当地表达出原文的语气、情态、语态、时态及所强调的重点。"简练"就是要求译文尽可能简短、精练，没有冗词废字，在明确、通顺的基础上力求简洁明快、精练流畅。

科技英语翻译大致可分为理解、表达、校正三个阶段。

1. 理解阶段

透彻理解原文是确切表达的前提。理解原文必须从整体出发，不能孤立地看待一词一句，因为每种语言几乎都存在着"一词多义"的现象。同样一个词或词组，在不同的上下文搭配中，在不同的语法结构中就可能有不同的意义，一个词、一个词组脱离上下文是不能被正确理解的。因此，译者首先应该结合上下文，通过对词义的选择、语法的分析，彻底弄清楚原文的内容和逻辑关系。通常情况下，理解是第一位的，表达是第二位的。当然，即使理解了原文，但不能用准确的汉语表达出来，也无法达到忠实表达原文思想内容的目的。为了透彻理解和准确表达，应注意以下几个方面。

（1）结合上下文，推敲词义　理解词义必须通过原文的上下文来进行，只有结合上下文才能了解单词在某一特定语言环境中的确切意义，否则翻译时容易出错。例如：

Various speeds may be obtained by the use of large and small pulleys.

错误译法：利用大小滑轮可以获得不同的转速。

正确译法：利用大小带轮可以获得不同的转速。

"pulleys"一般表示"滑轮"，但影响机器转速的零件名称应为"带轮"，而不是"滑轮"。

（2）辨明语法，弄清关系　科技英语的特点之一是句子长且语法结构复杂，因此，根据原文的句子结构弄清每句话的语法关系对正确理解原文具有重要意义。例如：

Intense light and heat in the open contrasted with the coolness of shaded avenues and the interiors of building.

错误译法：强烈的光线和露天场所的炎热，同林荫道上的凉爽和建筑内部形成了对比。

正确译法：露天场所的强烈光线和酷热，同林荫道上和建筑物内部的凉爽形成了对比。

（3）理解原文所涉及的事物　对科技英语的翻译，不能单靠语法关系来理解，还必须从逻辑意义和专业内容上来判断，应特别注意某些特有事物、典故和专业术语所表示的概念。例如：

Do you know that the bee navigates by polarized light and the fly controls its flight by its back wings?

错误译法：你知道蜜蜂借助极光导航，而苍蝇用后翅控制飞行吗？

正确译法：你知道蜜蜂借助偏振光导航，而苍蝇用后翅控制飞行吗？

2. 表达阶段

表达就是要寻找和选择恰当的词汇，把已经理解了的原文内容重新叙述出来。表达的好坏一般取决于对原文的理解深度和对目标翻译语言的掌握程度。理解正确并不意味着表达一定正确。

在理解阶段必须"钻进去"，把原文内容吃透；那么在表达阶段就必须"跳出来"，不受原文形式的束缚，要放开思路，按照汉语的规律和表达习惯遣词造句。表达阶段最重要的是翻译方法的选择，也就是如何"跳出来"的问题，这也是科技英语翻译的技巧问题，而科技英语翻译的创造性也表现在这方面。下面介绍几种基本的翻译方法。

（1）直译　是指译文采取原文的表现手法，既忠实于原文内容，又考虑原文形式。也就是说，在译文条件许可时，按照字面意思进行翻译，但直译绝不是逐词死译。例如：

Industrial regions of the world suffer much more acidic fall out than they did before Industrial Revolution.

直译：世界上的工业化地区现在遭受的酸性回落物的危害，比工业革命前要大得多。

（2）意译　是指在译文中用创新的表现手法来表现原文的逻辑内容和形象内容。当使用直译法不能使译文准确通顺时，就要使用意译。例如：

Mankind has always reverenced what Tennyson called "the useful trouble of the rain".

直译：人类一直很推崇坦尼森所说的"雨有用的麻烦"。

意译：人类一直很推崇坦尼森所说的这句话，"雨既有用，又带来麻烦"。

（3）音译　有些科学术语随着历史发展和技术进步已为人们所熟悉或掌握，翻译时往往由音译转为意译，或音译与意译同时采用。例如：moto 可译为"马达、电动机"，modern 可译为"摩登的、现代的"，microphone 可译为"麦克风、扩音器"，engine 可译为"引擎、发动机"，vitamin 可译为"维他命、维生素"。计量单位名称一般用音译，有些在翻译的时候还可以简化。例如：hertz 可译为"赫兹（赫）"，newton 可译为"牛顿（牛）"，pascal 可译为"帕斯卡（帕）"，watt 可译为"瓦特（瓦）"，volt 可译为"伏特（伏）"，ohm 可译为"欧姆（欧）"，lux 可译为"勒克斯（勒）"，bit 可译为"比特位（位）"，calorie 可译为"卡路里（卡）"等。

（4）形译　英语原文用字母表示事物外形时，汉语译文也可以按事物形状进行表达。例如：T-square 可译为"丁字尺"，I-steel 可译为"工字钢"，cross-road 可译为"十字路口"，A-frame 可译为"人字架"，V-slot 可译为"V 形槽"，Y-connection 可译为"星形联结"，X-ray可译为"X 射线"等。

在翻译实践中，应根据最能忠实、通顺地表达原文含义的原则，灵活机动地选用或交叉使用这四种翻译方法。翻译时应考虑到原作的整体性，最好以"段"而不是"句"为单位进行翻译，这不仅有利于辨别词义，而且有利于明晰句与句的衔接、段与段的联系，使译文不致成为一个个孤立句子的堆积。

3. 校正阶段

校正阶段是理解和表达的进一步深化，是使译文符合标准的一个必不可少的阶段，是对原文内容的进一步核实和对译文的进一步推敲。校正对于科技文献的翻译来说尤为重要，因为科技文章要求高度精确，公式、数据较多，稍一疏忽就可能造成严重的损失。理解和表达都不是一次完成的，而是逐步深入，最后才能完全理解和准确表达原文所反映的客观现实。因此，译文只有经过再三校正才能最终定稿。

Dialogue Exercise

Background：Li Lei and Han Meimei are going to visit an automatic factory.

Li Lei：Hello, Meimei. what about paying a visit to a "future" factory?

Han Meimei：It's a good idea. But when shall we go?

Unit 2
Virtual Reality

Li Lei: I think we'd better go there this Saturday. Are you free this Saturday?

Han Meimei: I think Saturday would be OK. But I have to check with my plan and see if I can finish my work for this week.

(Li Lei and Han Meimei are now in an automatic factory)

Han Meimei: Li Lei, look here at the assembly line. All the assembly work is done automatically with nobody working inside.

Li Lei: Yes, that's true. Everything here is done automatically. Even the quality control is done automatically.

Li Lei (**To the guide**): Excuse me, Ms. Why is there no light in your workshop? Do the machines work without light?

Guide: Yes. Most machines here don't need light for their work. They don't need any comfortable temperatures, either.

Han Meimei: But how could all this be made possible?

Guide: Well, you know. It's the computer that has made it possible.

Famous Quotes

Knowledge makes humble, ignorance makes proud.
博学使人谦逊，无知使人骄傲。

Unit 3　Internet of Things

知识目标：

1. 了解物联网技术的发展历程。
2. 掌握物联网技术的相关专业术语。
3. 掌握科技文章常见的篇章结构。

能力目标：

1. 能对物联网技术的专业术语进行中英互译。
2. 能对物联网技术相关英文资料进行阅读和翻译。
3. 能阅读分析典型的科技文章。

Reading Material

INTRODUCTION: The Internet of Things (*IOT*) is the Internet that connects things. The core and foundation of the Internet of things technology is still Internet technology. The application of the Internet of things technology extends between any item and object, for information exchange and communication.

The basic idea of the IOT is that virtually every physical thing in this world can also become a computer that is connected to the Internet. To be more accurate, things do not turn into computers, but they can feature tiny computers. [1] When they do so, they are often called smart things, because they can act smarter than things that have not been *tagged*.

Look around you for a second and count the number of electronic devices, machines and gadgets. All of them—light bulbs, cars, TVs, digital cameras, *refrigerators*, *stereos*, beds—will be connected to the Internet over the next 10 years (Fig. 3-1).

This is the potential of the "Internet of Things": billions and billions of devices and their components connected to one another via the Internet. The Internet of Things will radically alter our world through "smart" connectivity, save time and resources, and provide opportunities for innovation and

Fig. 3-1 IOT application

economic growth.

The basic building block of the Internet of Things is machine-to-machine (*M2M*) communication, devices equipped to communicate without the *intervention* of humans. The things of Internet have some key techniques:

1) *RFID*, sensors and so on could take information of things at anywhere and anytime.

2) The mixture of Internet could transmit the information accurately and quickly.

3) IPv6 address. In the Internet of things, everything should have a unique address. It makes the Internet of things possible. [2]

We should take a long term view of the Internet of things. The Internet of things is a new industry. It will bring many conveniences for our daily life. The development of the Internet of things is not only the development of the industry, but also the improvement of people's life.

Now we are during the *formation* stage of development. As the expansion of the use of the Internet of things, it will become mature and drive the development of the chain of the industry. In 3 to 5 years, there should be a mature application standard and technology standard of the industry. With the expansion of the industrial scale and the active of the application of the *transducer* technology, the standard system will be *nascent* framework. In 5 to 10years, the business model of the things of Internet will be more active. The market is going mature and standard of industry will be spread quickly and be widely recognized (Fig. 3-2).

It's no doubt that Chinese government has realized the importance of the Internet of things. But now we are during the formation stage of development. Government should support and apply the technology in some area like the government publication, *security* management, energy saving and environment protection.

The Internet of things is to combine various information sensing devices with Internet formed a huge network. Its purpose is to let all the goods and network linked together and easy to identify and management. The Internet of things is not the fantasy of technology, but yet another technological revolution. Industry experts believe that things can improve economic efficiency on the one hand, sig-

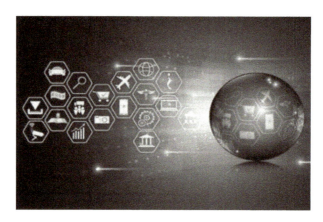

Fig. 3-2 The IOT for tomorrow

nificant cost savings; the other hand, the global economic recovery for the provision of technical power. [3] So we have reasons to believe that our future will be changed because of the Internet of things.

Words and Expressions

tag [tæg] vt. 标签；加标签于
refrigerator [rɪˈfrɪdʒəreɪtə(r)] n. 冰箱
stereo [ˈsteriəʊ] n. 立体声；立体声音响（器材）
intervention [ˌɪntəˈvenʃn] n. 介入，干预
formation [fɔːˈmeɪʃn] n. 形成；结构
transducer [trænzˈdjuːsər] n. 传感器；变频器
nascent [ˈnæsnt] adj. 初期的；初生的
security [sɪˈkjʊərəti] adj. 安全的；保密的
IOT 物联网
M2M 机器通信
RFID (Radio Frequency Identification) 射频识别技术

Special Difficulties

1. To be more accurate, things do not turn into computers, but they can feature tiny computers.
to be more accurate 意为"更准确地说。"
本句可译为：更准确地说，事物可具有微型计算机的特点，但不会变成计算机。

2. IPv6 address. In the Internet of things, everything should have a unique address. It makes the Internet of things possible.
IPv6 是 "Internet Protocol Version 6" 的缩写，其中 Internet Protocol 意为"互联网协议"。
本句可译为：IPv6 地址。在物联网中，每个物体都应该有一个唯一的地址。这使得物

联网成为可能。

3. Industry experts believe that things can improve economic efficiency on the one hand, significant cost savings; the other hand, the global economic recovery for the provision of technical power.

on the one hand..., the other hand 意为"一方面……，另一方面……"。

本句可译为：业内专家认为，物联网一方面可以提高经济效益，大大节约成本；另一方面，可以为全球经济复苏提供技术动力。

Learn and Practice

1. Mark the following statements with T (true) or F (false) according to the text.

1) The things of internet have some key techniques: RFRD, sensors. (   )

2) With the expansion of the industrial scale and the active of the application of the transducer technology, the standard system will be nascent framework. (   )

3) In the internet of things, everything should have a single address. (   )

2. Choose the best choices according to the text.

1) The internet of things is to combine various information sensing devices with Internet formed a huge (   ).

　　A. network　　　　　B. net　　　　　　C. framework

2) Look around you for a second and count the number of (   ) devices, machines and gadgets.

　　A. electronics　　　　B. electronical　　　C. electronic

3) The Internet of Things will (   ) alter our world through "smart" connectivity, save time and resources, and provide opportunities for innovation and economic growth.

　　A. radical　　　　　B. radically　　　　C. finally

4) Government should support and apply the (   ) in some area like the publish government, security management, energy saving and environment protection.

　　A. manpower　　　　B. money　　　　　C. technology

3. Translate the following passage into Chinese.

It's no doubt that Chinese government has realized the importance of the Internet of things. But now we are during the formation stage of development. Government should support the industry and apply the technology in some area like the publish government, security management, energy saving and environment protection.

Extensive Reading

Data Warehouse and Data Mining

Data warehouse is a database that *consolidates* data extracted from various production and *operational* systems into one large database that can be used for management reporting and analysis (Fig. 3-3). The data from the organization's core transaction processing system are reorganized and combined with other information, including historical data so that they can be used for management decisions and analysis.

A data warehouse is becoming more of a necessity than an *accessory* for a *progressive*, competitive, and focused organization. It provides the right foundation for building decision support and *executive* information system tools that are often built to measure and provide a feet for how well an organization is progressing toward its goals.

Fig. 3-3 Data Warehouse

In most cases, the data in the data warehouse can be used for reporting, they cannot be updated so that the *performance* of the company's underlying operational system is not affected.[2] The focus on problem solving describes some of the *benefits* companies have obtained by using data warehouses.

Data warehouses often contain capabilities to *remodel* the data. A relational database allows views of data into two dimensions. A multidimensional view of data lets users look at data in more than two dimensions.

Data mining is about analyzing data and finding hidden patterns using automatic or semiautomatic means (Fig. 3-4). Data mining provides a lot of business value for enterprises. Such as increasing competition, customer segmentation, churn analysis, cross-selling, sales forecast, fraud detection, risk management, and so on.

During the past decade, large volumes of data have been *accumulated* and stored in databases. Much of this data come from business software, such as financial applications, *Enterprise Resource Planning* (ERP), *Customer Relationship Management* (CRM), and Web logs. The result of this data collection is that organizations have become data-rich and knowledge-poor. The collection of data has become limited. The main purpose of data mining is to extract patterns from the data at hand, increase its intrinsic value and transfer the data to knowledge.[3]

Data mining applies *algorithms*, such as decision tree, clustering, *association*, time series, and

Fig. 3-4   Data Mining

so on, to a data set and analyses its contents. This analysis produces patterns, which can be explored for valuable information. Depending on the underlying algorithm, these patterns can be in the form of trees, rules, clusters, or simply a set of *mathematical* formulas. The information found in the patterns can be used for reporting, as a guide to marketing strategies, and, most importantly, for prediction.

Words and Expressions

consolidate [kən'sɒlɪdeɪt] v. 巩固，使固定；联合
operational ['ɒpə'reɪsənl] adj. 操作的；运作的
accessory [ək'sesəri] n. 配件，附件
progressive [prə'gresɪv] adj. 进步的；先进的
executive [ɪg'zekjətɪv] adj. 行政的；经营的；执行的
performance [pə'fɔːməns] n. 性能；表演；表现
benefit ['benɪfɪt] n. 利益，好处；救济金
remodel [ˌriː'mɒdl] v. 改造；改变；改型
accumulate [ə'kjuːmjəleɪt] v. 积累，积聚，堆积
algorithm ['ælgərɪðəm] n. 算法，运算法则
association [əˌsəʊʃi'eɪʃn] n. 协会，联盟；联合
mathematical [ˌmæθə'mætɪkl] adj. 数学的，数学上的；精确的
Enterprise Resource Planning   企业资源管理
Customer Relationship Management 客户关系管理

Special Difficulties

1. Data warehouse is a database that consolidates data extracted from various production and operational systems into one large database that can be used for management reporting and analysis.

be used for 意为"被用于……"。

本句可译为：数据仓库是一种数据库，它将从各种生产和运营系统中提取的数据合并到一个大型数据库中，可用于管理状况的报告和分析。

2. In most cases, the data in the data warehouse can be used for reporting, they cannot be updated so that the performance of the company's underlying operational system is not affected.

in most cases 意为"在大多数情况下"。

本句可译为：在大多数情况下，数据仓库中的数据可用来进行报告，不可进行更新，所以公司的基础运营系统的性能不会受到影响。

3. The main purpose of data mining is to extract patterns from the data at hand, increase its intrinsic value and transfer the data to knowledge.

the main purpose of... 意为"……的主要目的"。

本句可译为：数据挖掘的主要目的是从现有数据中提取模式，增加数据的自身价值，并将这些数据转变为知识。

Learn and Practice

1. Mark the following statements with T (true) or F (false) according to the text.

1) Data mining provides the right foundation for building decision support and executive information system tools. (   )

2) During the past decade, small volumes of data have been accumulated and stored in databases. (   )

3) In most cases, the data in the data warehouse can be used for reporting. (   )

4) A multidimensional view of data lets users look at data in more than two dimensions. (   )

2. Choose the best choices according to the text.

1) The focus on problem solving describes some of the (   ) companies have obtained by using data warehouses.

   A. strength   B. benefits   C. benefit

2) Data warehouses often contain capabilities (   ) remodel the data.

   A. with   B. of   C. to

3) The result of this data collection in that organizations have (   ) data-rich and knowledge-poor.

   A. reach   B. become   C. attain

3. Translate the following passage into Chinese.

Data warehouses often contain capabilities to remodel the data. A relational database allows views of data into two dimensions. A multidimensional view of data lets users look at data in more than two dimensions.

Knowledge Link

科技文章的篇章结构

一、一般结构

在阅读一篇完整的科技英语文章时，仔细分析一下这类文章的篇章结构，可以发现科技英语文章一般由以下几部分组成：

1. 标题 Title
2. 目录 Contents
3. 摘要 Abstract（包含关键词 Key Words）
4. 引言 Introduction，前言 Preface
5. 正文 Body
6. 结论与建议 Conclusions and Suggestions
7. 总结 Summary
8. 致谢 Acknowledgement
9. 注释 Notes
10. 参考文献 References
11. 附录 Appendix

但对于一般的科技英语文章来说，不一定需要上述齐全的文体结构，可以进行取舍。常见科技英语文章的文体结构如下：

1. 标题 Title
2. 摘要 Abstract（包含关键词 Key Words）
3. 引言（概述） Introduction（General Description）
4. 正文 Body
5. 结论与建议 Conclusions and Suggestions
6. 参考文献 References

二、语言特点

1. 标题

科技英语文章的标题往往表达文章的中心主旨，文章的标题反映了作者所研究的主要内容，或作者所描述的科学真理和事实，或作者要说明的科学实验过程和产品使用说明等，其语言特征如下。

1）关于……的研究、探讨、调查、介绍、分析、描述等说明性内容，其表达方式如下：

research/study/probe on...

introduction/brief introduction to...

investigation/survey/analysis/description of...

2）说明科学真理和事实、试验过程、产品说明等，一般采用名词性词组、现在分词的形式，示例如下：

The Miracle Chip, Machine Tools, Computer in the Future, Robots for Tomorrow, Digital Age, Information Highway, The Jet Engine, Operating System, Cloning Technology。

2. 摘要

摘要反映一篇科技英语文章论述的主要内容、思想方法、重要观点和结论等，一般摘要的长度为150~200字，主要由三部分组成，即主题句、支持句和结论句，其主要语言特征如下。

（1）主题句常用句型示例

1）The purpose of this paper is...

2）The primary goal of this research is...

3）The overall objective of this study is...

4）In this paper, we aim at...

5）Our goal has been to provide...

6）The chief aim of the present work is to investigate the facts that...

7）The main objective of our study has been to obtain some results...

（2）支持句常用句型示例

1）The method used in our study is known as...

2）The technique we applied is referred to as...

3）The procedure they followed can be briefly described as...

4）The approach adopted extensively is called...

5）Detailed information has been acquired by the authors using...

6）The research has recorded valuable data using the newly-developed method...

7）This is a working theory which is based on the idea that...

8）The fundamental feature of this theory is as follows...

9）The theory is characterized by…

10）The experiment consisted of three steps, which are described in…

11）The test equipment that was used consisted of...

12）Included in the experiment were...

（3）结论句常用句型示例

1）In conclusion, we state that...

2）In summing up, it may be stated that...

3）It is concluded that...

4）The results of the experiment indicate that...

5）The studies we have performed showed that...

6）The pioneer studies that the authors attempted have indicated in...

7）We carried out several studies that have demonstrated that...

8）This fruitful work gives explanation of...

9）The author's pioneer work has contributed to our present understanding of...

10）The research work has brought about a discovery of...

11）These findings of the research have led the author to the conclusion that...

12）The data obtained appear to be very similar to those reported earlier by…

3. 引言

引言是对全文的综合和概述，包括研究的背景、目的及意义，用于向读者介绍文章的思想和内容，其常用句型示例如下：

1）Over the past several decades...

2）Somebody reported...

3）The previous work on... has indicated that...

4）Recent experiments by... have suggested...

5）Several researchers have theoretically investigated...

6）In most studies of... has been emphasized with attention being given to…

7）Industrial use of... is becoming increasingly common…

8）There have been a few studies highlighting...

9）It is well known that...

4. 正文

正文是科技英语文章的主体，是科学分析和试验论证的过程反映，经常使用各种图表、公式论证作者的观点。图表相关词句的翻译如下：

1）表：Table.

2）图：Figure/Diagram/Graph/View/Flow Diagram/Chart/Frame Figure，and so on.

3）公式、算式、方程式：Formula/Equation.

4）如图 X 所示，如表 Y 所示：As it is shown in Fig. X./As it is shown in Table Y.

5. 结论

结论是对全文的总结，是经过科学分析、研究而得出的结论，其常用句型示例如下：

1）The following conclusions can be drawn from...

2）It can be concluded that...

3）We may conclude that...

4）We come to the conclusion that...

5）It is generally accepted (believed, held, acknowledged) that...

6）We think (consider, believe, feel) that...

7）It is advantageous to do...

8）It should be realized (emphasized, stressed, noted, pointed out) that...

9）It is suggested (proposed, recommended, desirable) that...

10）It would be better (helpful, advisable) that...

6. 致谢

致谢是作者在完成文章之后，表达对帮助了自己的人士的感谢，其常用句型示例如下：

1）I am thankful to sb. for sth.

2）I am grateful to sb. for sth.

3）I am deeply indebted to sb. for sth.

4）I would like to thank sb. for sth.

5）Thanks are due to sb. for sth.

6）The author wishes to express his sincere appreciation to sb. for sth.

7）The author wishes to acknowledge sb.

8）The author wishes to express his gratitude to sb. for sth.

7. 参考文献

在文章最后应将参考过的主要文献一一列出，表示对他人成果的尊重及作者的写作依据。

Dialogue Exercise

Background：Han Meimei now in a factory, she is talking to Li Lei who specialize in NC machine production.

Han Meimei：Good morning, Li Lei.

Li Lei：Hi, Meimei.

Han Meimei：What are you doing now?

Li Lei：I am turning a work-piece.

Han Meimei：Is this lathe different from that lathe?

Li Lei：Yes, It is.

Han Meimei：Where are the distinctions about them?

Li Lei：This is an NC lathe, and the other is a conventional lathe.

Han Meimei：What is NC?

Li Lei：NC refers to numerical control, and controlling a machine tool by means of a prepared program.

Han Meimei：What does the NC machine tool system contain?

Li Lei：It contains the machine control unit and the machine tool itself.

Han Meimei：Do you like to machine a work-piece on NC machine tools or conventional machine tools.

Li Lei：I would prefer to use NC machine tools very much.

Han Meimei：Why do you think that?

Li Lei：Because it has some advantages such as increased productivity, reduced noise, improved operator's safety and product quality, and so on.

Han Meimei: It sounds quite well.

Li Lei: You will like it in the future, too.

Han Meimei: I think so. Thanks a lot.

Li Lei: You are welcome.

Famous Quotes

Whatever is worth doing is worth doing well.
凡是值得做的事,就值得做好。

Unit 4　3D Printing

知识目标：

1. 了解3D打印技术的发展历程和工作原理。
2. 掌握3D打印技术的相关专业术语。
3. 掌握常用科技英语的句型特点。

能力目标：

1. 能对3D打印技术的专业术语进行中英互译。
2. 能对3D打印技术相关英文资料进行阅读和翻译。
3. 能正确分析科技英语中的长难句。

Reading Material

INTRODUCTION： The industrial revolution of the late 18th century made possible the mass production of goods, thereby creating economies of scale which changed the economy and society in ways that nobody could have imagined at the time.[1] Now a new manufacturing technology has emerged which does the opposite. Three-dimensional printing makes it as cheap to create single items as it is to produce thousands and thus undermines economies of scale.[2] It may have as profound an impact on the world as the coming of the factory did.

The term "3D printing" was coined at MIT in 1995 when graduate students Jim Bredt and Tim Anderson *modified* an inkjet printer to *extrude* a binding solution onto a bed of powder, rather than ink onto paper (Fig. 4-1). The ensuing *patent* led to the creation of modern 3D printing companies Z Corporation (founded by Bredt and Anderson) and ExOne.

It works like this. First you call up a blueprint (*digital model*) on your computer screen and tinker with its shape and color where necessary. Then you press "print". A machine nearby whirrs into life and builds up the object gradually, either by depositing material from a nozzle, or by selectively solidifying a thin layer of plastic or metal dust using tiny drops of glue or a tightly focused *beam*. [3] Products are thus built up by progressively adding material, one layer at a time, hence the technology's other name, additive manufacturing. Eventually the object such as a part for your car, a lampshade, a violin—pops out (Fig. 4-2). The beauty of the technology is that it does not need to happen in a factory. Small items can be made by a machine like a desktop printer, in the corner of an office, a shop or even a house; big items like bicycle *frames*, panels for cars, aircraft parts need a larger machine, and a bit more space.

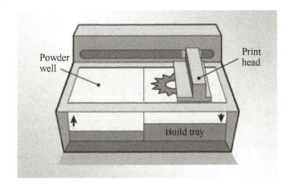

Fig. 4-1   The principle model of 3D printer

Fig. 4-2   The 3D printer

The additive approach to manufacturing has several big advantages over the conventional one. It cuts costs by getting rid of production lines. It reduces waste enormously, requiring as little as one-tenth of the amount of material. It allows the creation of parts in shapes that conventional techniques cannot achieve, resulting in new, much more efficient designs in aircraft wings or heat exchangers, for example. It enables the production of a single item quickly and cheaply.

At the moment the process is possible only with certain materials (plastics, *resins* and metals) and with a precision of around a tenth of a millimeter. As with computing in the late 1970s, it is currently the preserve of *hobbyists* and workers in a few *academic* and industrial *niches*. But like computing before it, 3D printing is spreading fast as the technology improves and costs fall. A basic 3D printer, also known as a fabricator or "fabber", [4] now costs less than a laser printer did in 1985.

The technology will have implications not just for the distribution of capital and jobs, but also for intellectual-property (IP) rules. When objects can be described in a digital file, they become much easier to copy and distribute. Just ask the music industry. When the blueprints for a new toy, or a designer shoe, escape onto the internet, the chances that the owner of the IP will lose out are greater.

Just as nobody could have predicted the impact of the steam engine in 1750, or the printing press in 1450, or the transistor in 1950, it is impossible to foresee the long-term impact of 3D printing. But the technology is coming, and it is likely to disrupt every field it touches. Companies, regulators and entrepreneurs should start thinking about it now. One thing, at least, seems clear: although 3D printing will create winners and losers in the short term, in the long run it will expand the realm of industry.

Words and Expressions

modify ['mɒdɪfaɪ] v. 修改，更改
extrude [ɪk'struːd] v. （被）挤压出；喷出
patent ['pætnt] n. 专利；专利品
beam [biːm] n. 梁；光线；（电波的）波束
frames [freɪmz] n. 框架；边框
resin ['rezɪn] n. 树脂；松香
hobbyist ['hɒbiɪst] n. 爱好者
academic [ˌækə'demɪk] adj. 学院的，大学的；学会的
niches ['nɪtʃiz] n. 合适的位置
3D printing 3D 打印
digital model 数字模型

Special Difficulties

1. The industrial revolution of the late 18th century made possible the mass production of goods, thereby creating economies of scale which changed the economy and society in ways that nobody could have imagined at the time.

thereby 意为"由此，从而"。

本句可译为：18 世纪后期的工业革命使得商品的大规模生产成为可能，从而创造了规模经济，以当时人们所想象不到的方式影响了经济和社会的发展。

2. Three-dimensional printing makes it as cheap to create single items as it is to produce thousands and thus undermines economies of scale.

as... as... 意为"……和……一样"。

本句可译为：3D 打印使得单件生产和规模生产的成本一样低，从而削弱了规模经济。

3. A machine nearby whirrs into life and builds up the object gradually, either by depositing material from a nozzle, or by selectively solidifying a thin layer of plastic or metal dust using tiny drops of glue or a tightly focused beam.

whirrs into life 意为"带着轻微的响声进入工作状态"。再比如 Suddenly the hall burst into life，可译为：大厅突然活跃起来；The fire flared into life，可译为：火旺了起来。

本句可译为：摆在附近的机器就会开始工作，发出轻微的响声，通过喷嘴注入并沉积材料，或通过胶合剂或者激光选择性地固化薄塑料层或金属粉末层。

4. fabber

一款免费的开源的快速成型机的名字。任何人都可以到其官方网站下载制做快速成型机的资料。

Unit 4
3D Printing

Learn and Practice

1. Mark the following statements with T (true) or F (false) according to the text.

1) The industrial revolution of the late 18th century made the mass production of goods impossible. (　)

2) The process of 3D printing is possible only with plastics. (　)

3) We have predicted the impact of 3D printing. (　)

2. Choose the best choices according to the text.

1) It works like this. First you call up a blueprint, a (　) model on your computer screen and tinker with its shape and color where necessary.
   A. digital   B. code   C. coded

2) The ensuing patent (　) to the creation of modern 3D printing companies Z Corporation (founded by Bredt and Anderson) and ExOne.
   A. led   B. used   C. made

3) Products are thus built up by progressively adding material, one layer at a time, hence the technology's other name, (　) manufacturing.
   A. additive   B. cumulative   C. additives

4) The term "3D printing" was coined at (　) in 1995.
   A. MIT   B. Harvard   C. Yale

5) At the moment the process is possible only with certain materials (plastics, resins and metals) and with a precision of around a (　) of a millimeter.
   A. 10th   B. ten   C. 100th

3. Translate the following phrases into Chinese.

1) Additive manufacturing.
2) Intellectual-property (IP).
3) Long-term impact.

Extensive Reading

Advanced Manufacturing Technology

Progress in human society has been accomplished by the creation of new technologies. The last few years have *witnessed unparalleled* changes throughout the world. [1] Rapid changes in the markets demand *drastically* shortened product life cycles and high-quality products at competitive prices. Customers now prefer a large variety of products. This *phenomenon* has *inspired* manufacturing firms to look for progressive computerized automation in various processes. Thus mass production is being replaced by low-volume, high-variety production. Manufacturing firms have recognized the importance of flexibility in the manufacturing system to meet the challenges posed by the *pluralistic* market. The concept of flexibility in manufacturing systems has attained significant importance in meeting the challenges for a variety of products of shorter *lead-times*, together with higher productivity and quality. The flexibility is the underlying concept behind the transition from traditional methods of production to the more automated and *integrated* methods (Fig. 4-3). They stress that firms *implementing* automation projects should *prioritize* their needs for different flexibilities for long-range strategic perspectives.

Fig. 4-3　AMT in production

Numerous definitions of AMT exist. For example, Baldwin (1995) defines AMT as a group of integrated hardware-based and software-based technologies, which if properly implemented, monitored, and *evaluated*, will lead to improving the *efficiency* and effectiveness of the firm in manufacturing a product or providing a service. AMT, defined broadly, is a total socio-technical system where the adopted *methodology* defines the *incorporated* level of technology. AMT employs a family of computer aided manufacturing (CAM), flexible manufacturing systems (FMS), manufacturing resource planning (MRP), automated material handling systems, robotics, computer numerically controlled (CNC) machines (Fig. 4-4), computer-integrated manufacturing (CIM) systems, *optimized* production technology (OPT), [2] and just-in-time (JIT). [3] Although AMT places great *emphasis* on the use of technological innovation, management's role is significant since AMT systems require continual review and readjustment.

The properties inherent in Advanced Manufacturing Technology (AMT) create new opportunities for firms, and in particular small firms in the local context. [4] The capability of this technology to modify production specifications quickly and accurately means that firms can customize their products and attain economics of scope based on low volume and low cost production. While traditionally technology has been *perceived* merely as a tool in implementing business strategy, AMT has the potential to directly affect the firm's strategy choices. *To date*, AMT literature suggests that adoption of AMT offers firms the potential to pursue new innovative strategies.

Fig. 4-4   CNC production

Words and Expressions

witness ['wɪtnəs] v. 作证；表示
unparalleled [ʌn'pærəleld] adj. 无比的，无双的
drastically ['drɑːstɪklɪ] adv. 彻底地，激烈地
phenomenon [fə'nɒmɪnən] n. 现象；事件
inspired [ɪn'spaɪəd] v. 鼓舞；激励
pluralistic [ˌpluərə'lɪstɪk] adj. 兼职的；多元化的
integrated ['ɪntɪgreɪtɪd] adj. 完整的；整体的；结合的
implementing ['ɪmplɪməntɪŋ] v. 实现；执行；使生效
prioritize [praɪ'ɒrətaɪz] vt. 按重要性排列；优先处理
evaluate [ɪ'væljueɪt] v. 评价；求……的值（或数）
efficiency [ɪ'fɪʃnsi] n. 效率；能力
methodology [ˌmeθə'dɒlədʒɪ] n. 方法学，方法论
incorporated [ɪn'kɔːpəreɪtɪd] adj. 股份有限的，组成公司的
optimize ['ɒptɪmaɪz] adj. 最佳化的，（使）最优化的
emphasis ['emfəsɪs] n. 强调，突出
perceive [pə'siːv] v. 感觉；理解为
lead-times 交付周期；更换模具的时间
to date 到目前为止，迄今

Special Difficulties

1. Progress in human society has been accomplished by the creation of new technologies. The last few years have witnessed unparalleled changes throughout the world.

the last few years 意为"最近的几年",也可以用 the past few years(过去的几年),均表示完成时态,后面省略主语。

本句可译为:技术创新带动着人类社会的进步,近几年世界的变化更是日新异。

2. Optimized Production Technology (OPT).

最佳生产技术(OPT)是一种改善生产管理的技术,由以色列物理学家 Eli Goldratt 博士于 20 世纪 70 年代提出,是用于安排企业生产人力和物料调度的计划方法。

3. Just In Time (JIT).

准时制生产方式(JIT)又称无库存生产方式(stockless production)、零库存生产方式(zero inventories)、单件流生产方式(one-piece flow)或者超级市场生产方式(supermarket production)。

4. The properties inherent in Advanced Manufacturing Technology (AMT) create new opportunities for firms, and in particular small firms in the local context.

本句可译为:先进制造技术(AMT)的内在优势会为企业,特别是地区性小企业创造新的发展机会。

Learn and Practice

1. Mark the following statements with T (true) or F (false) according to the text.

1) Customers prefer a large variety of products nowadays. (   )

2) Most manufacturing firms have not recognized the importance of flexibility in the manufacturing system. (   )

3) There is only one definition of AMT. (   )

2. Translate the following passage into Chinese.

Owing to the intense global competition in manufacturing, manufacturers need to increase their level of competitiveness in the global market. Some manufacturing companies, therefore, are forced to undergo a period of transformation in order to compete more effectively. Under these circumstances, AMT is considered as a means of improving competitiveness.

Knowledge Link

科技英语的句型分析

一、科技英语的句型结构

科技英语文章中的句型和普通英语文章中的句型用法基本相同,只是长难句和复合句较多。但英语文章中的句子不论多长或者多么复杂,不考虑句子之间的内在联系,按照一定规律抓住句子主干并对句子成分进行分解,都可以将长句拆分成若干简单句。

一般来说,动词是英语句子的主干。动词不同,英语句子的主要成分就会不同,从而形成不同的句型。因此，我们要分析英语句子的结构，首先要弄清句子的类型。

1. 常见的三种基本句型

首先说明，以下部分中 S 指代 subject（主语），V 指代 verb（动词），O 指代 object（宾语），P 指代 predicative（表语），C 指代 complement（补语），L 指代 link verb（系动词），DO 指代 direct object（直接宾语），IO 指代 indirect object（间接宾语）。

(1) SV［主语 + 谓语（不及物动词）］结构

1) In a hydrogen atom, an electron whirls around the nucleus at a tremendous speed.

氢原子中，电子以极高的速度绕原子核旋转。

2) Electro magnetic waves can move through great distance.

电磁波可以传播很远的距离。

(2) SVO［主语 + 谓语（及物动词）+ 宾语］结构

这种句型又可分为三类，分别是"主语 + 谓语 + 宾语""主语 + 谓语 + 宾语 + 宾语补语""主语 + 谓语 + 间接宾语 + 直接宾语"。

1) Some computers can perform over a billion computations a second.

有些计算机每秒可以完成十亿次以上的运算。

2) During war, radar enables the bombers to find their targets at night.

发生战争时，雷达可以帮助轰炸机在黑夜里找到目标。

3) These new methods will make the electronic devices of the future quite small.

这些新方法将使未来的电子设备变得很小。

4) The turbojet and the turboprop give airplanes an even thrust.

涡轮喷气发动机和涡轮螺旋桨发动机给飞机提供平稳的推力。

5) This gives the work a good finish.

这使工件有良好的表面质量。

(3) SLC（主语 + 系动词 + 表语）

1) The laser is really amazingly simple in construction.

激光器在结构上是极其简单的。

2) At about 1300℃ the metal becomes plastic.

在大约1300℃时，金属会变成塑性体。

2. 并列句和复合句的基本类型

如果句子中出现不止一个主谓结构，就要用连词组成并列句或复合句。以两个分句组成一个句子为例，按照主谓结构安排不同有以下几种句型。

(1)（主语 + 谓语）+（连词 + 主语 + 谓语）

1) Most sonar sets send out sounds that are millions of times more powerful than a shout.

多数声呐装置发出的声音比喊叫声要强数百万倍。

2) A box resting on the floor has more than one force acting on it, but they do not produce a change in its position.

一个放在地板上的箱子有不止一个力作用于它，但是这些力不会使箱子的位置发生变化。

(2)（连词 + 主语 + 谓语）+（主语 + 谓语）

1) As load is added, the active component of the current increases.

加上负载后,有效电流增大。

2) As we know, there are two kinds of steel, carbon steel and alloy steel.

如我们所知,钢有两种类型:碳钢和合金钢。

(3) 主语 +(连词 + 主语 + 谓语)+ 谓语

1) An earth satellite, whether it is natural or artificial, is held in orbit by the balance of gravity and the satellite's inertia.

地球卫星,不管是自然的还是人造的,都靠地心引力与卫星惯性之间的平衡将其保持在轨道上。

2) The petrol you use to drive your car engine is fined product of crude oil.

驱动汽车发动机所用的汽油是由原油提炼出来的。

(4)(连词 + 主语 + 谓语)+ 谓语

1) That matter consists of atoms is known to all of us.

我们都知道物质由原子组成。

2) Whether life may exist on the Moon or Mars remains a mystery.

月球或火星上是否存在生命还是个谜。

二、科技英语的句型特点

1. 多重复合句

科技英语文章常常使用这种句型,以便能严谨地表达复杂的内容,如果把一句话分成几个独立的句子,这样就有可能影响句子之间的密切联系。所以说,多重复合句是最能体现科技英语文体特点的一种句型。文章的论述性越强,多重复合句用得越多,句子也越长。多重复合句的分句之间有两种基本关系,一种是并列关系,另一种是主从关系;但是以主从关系为主。这两种关系有时也会同时出现在一个句子中。

科技英语文章中有的句子可能很长,遇到这种句子时,不要"眉毛胡子一把抓",而是要进行语法分析。语法分析主要从两点入手,第一是找出谓语(谓语的形式比较明显,容易发现),然后找出对应的主语。英语句子不像汉语那样经常省略主语,而是由主语和谓语构成句子的主干。第二是找出连接词,英语和汉语的另一个不同之处是汉语句子分句时常常没有连接词,而英语句子的分句之间一般都有连接词,找出了连接词就找到了分句间的界限和它们之间的关系。这里说的连接词是广义的,包括连接代词、连接关系代词、关系副词等。当然有些连接词是通用的,读者还需要根据分句成分进行具体分析。

2. 被动语态

在科技英语文章中大量存在被动语态,这是因为科技英语文章中往往不需要明确动作的执行者,一般采用被动语态来表达;但翻译时可采用主动语态。同时,科技英语文章之所以多用被动语态,也是为了强调所论述的客观事物。

1) No work can be done without energy.

没有能量就不能做功。

2) All sorts of necessaries of life can be made of plastics.

各种生活必需品都能用塑料来制造。

Unit 4
3D Printing

3. 非谓语形式

科技英语文章中的句子一般比较简单，通常只用一个谓语动词。如果有几个动作，就必须选出主要谓语动词，而将其余动词表达为非谓语动词形式，才符合英语语法要求。非谓语动词有三种：动名词、分词（包括现在分词和过去分词）和不定式。非谓语动词在科技英语文章的简单句中使用得非常频繁。示例分析如下：

（1）To be a true professional requires lifelong learning.

本句可译为：要成为一名真正的专业人士，需要终生学习。

这个句子中有"成为""需要"和"学习"三个动词。可以看出，"需要"（require）是主要谓语动词，其余两个动作，"成为"采用不定式形式"to be"，而"学习"采用动名词形式"learning"。

（2）Heating water does not change its chemical composition.

本句可译为：加热水并不会改变水的化学成分。

这句中有"加热"和"改变"两个动词，本句的处理方式是将"改变"（change）用作主要谓语，而将"加热"处理为动名词"heating"，连同其宾语 water 作为本句主语。

（3）Matter is anything having weight and occupying space.

本句可译为：任何具有重量并占有空间的东西都是物质。

这句包含"是""具有"和"占有"三个动作，将"是"处理成主要谓语（系动词）"is"，而"具有"和"占有"处理为现在分词"having"和"occupying"，同它们的宾语"weight"和"space"分别构成现在分词短语，作为修饰名词"anything"的后置定语。

4. 词性转换

英语单词有不少多义词、多性词，即既可以是名词，在经过词性转换后又可作动词、形容词、介词或者副词，表达另一个意思。如果不仔细阅读分析，在翻译的过程中往往会出现错误。词性转换在科技英语文章中屡见不鲜，几乎每个技术名词都可转换为同义的形容词，如capital goods（生产资料），space rocket（宇宙火箭）。词性转换增加了科技英语的灵活性和表现力，读者必须从上下文判明用词在句中是何种词性、含义如何，才能对全句做到正确无误的理解。

词性转换请读者参考本书科技英语词汇构成的相关内容（Unit 5～Unit 6）。

三、句子之间的逻辑关系

科技英语文章由句子、句群或段落组成，文章的各组成部分存在一定的逻辑关系。作者为了准确表达一个科学真理、一个科学事实或一个试验过程，往往采用一些虚词来表达逻辑关系，如连词、副词、介词短语、不定式短语等，这些词多出现在句首或句中，较少出现在句尾。句子较长时，各分句也有自己的逻辑关系词语，以更好地表达句子内部之间的联系，有助于准确表达作者的思想和意图。表示逻辑关系的示例词如下：

1）表示列举：firstly, secondly, finally, for one thing...(and) for another (thing), in the first place, to begin with, initially, next, lastly, on the other hand, etc.

2）表示增补：and, and also, in addition (to), furthermore, moreover, what is more, as well as, etc.

3）表示转折或对比：but, yet, nevertheless, instead, in fact, on the contrary, on the one hand... on the other (hand), however, as a matter of fact, etc.

4) 表示原因或结果：hence, therefore, thus, consequently, because of this, for this reason, in consequence, on account of this, as a result, etc.

5) 表示解释：that is to say, namely, for example, for instance, such as, in other words, etc.

6) 表示总结或结论：(all) in all, in conclusion, in short, in a word, in brief, on the whole, to conclude, to sum up, the result of, apparently, seemingly, undoubtedly, etc.

Dialogue Exercise

Background: Li Lei who works for a company which produces 3D printer as a salesman, shows their workshop and the latest product to Han Meimei, a potential customer from another firm.

Han Meimei: It was very kind of you to give me a tour of the place. It gave me a good idea of your product range.

Li Lei: It's a pleasure to show our factory to our customers. What's your general impression, may I ask?

Han Meimei: Very impressive, indeed, especially the speed of your new 3D printer.

Li Lei: That's our latest product. We put it on the market just two months ago.

Han Meimei: The machine gives you an edge over your competitors, I guess.

Li Lei: Certainly. No one can match us as far as speed is concerned.

Han Meimei: Could you give me some brochures for that machine?

Li Lei: Right. Here is our sales catalog and literature.

Han Meimei: Thank you. I think we may be able to work together in the future.

> **Famous Quotes**
>
> **Every great advance in science has issued from a new audacity of imagination.**
> 科学上的每一项重大进步都来自一种新的大胆设想。

Unit 5　Industrial Robot

知识目标：

1. 掌握工业机器人的结构和工作原理。
2. 掌握工业机器人的应用领域和相关专业术语。
3. 掌握科技英语的构词方法。

能力目标：

1. 能对工业机器人的专业术语进行中英互译。
2. 能对工业机器人相关英文资料进行阅读和翻译。
3. 能正确理解科技英语中的专业术语。

Reading Material

INTRODUCTION：This paper mainly describes the robot definition, the development of robots and the industrial robot's kinematic structure. A robot is a re-programmable, multi-functional manipulator designed to move materials, parts, tools and special devices through a variety of programmed motions for the performance campaigns to control a variety of different tasks.

A robot is a reprogrammable, multi-functional *manipulator* designed to move materials, parts, tools and special devices through a variety of programmed motions for the performance *campaigns* to control a variety of different tasks (Fig 5-1). The industrial robot is a tool that is used in the manufacturing environment to increase productivity. It can perform jobs that might be hazardous to the human worker. One of the first industrial robots was used to replace the nuclear fuel *rods* in nuclear power plant. The industrial robot can also operate on the assembly line such as placing electronic components on a printed circuit board. Thus, the human worker can be relieved of the routine operation of this tedious task. Robots can also be programmed to defuse bombs, to serve the handicapped, and to perform functions in numerous applications in our society.

In contrast with the conventional view, *robotics* was not developed recently.[1] In fact, the first robots were created in the early 1960's in America. Unimation produced a robot arm in 1961 with the

Fig. 5-1 Industrial robot for automatic assembly

control unit sequence set by the operator. But due to the experimental nature of the work, a low profile was kept mainly to avoid adverse public reaction to the project. In 1974 Cincinnati Millicron was the first mini-computer controlled robot. However, in the same year, the IRB6 robot was introduced by the Swedish company ASEA. This robot has been marketed all over the world and is still in production today with the only significant improvements in the control counter electronic devices and software upgrades. Therefore, when the United States may be credited for establishing the technology for robotics, countries like Japan and Sweden have utilized it to a greater extent in industrial applications. Today current research efforts focus on creating a "smart" robot that can "see" "hear" "touch" and consequently make decisions. [2]

To be able to classify what order of design an industrial robot belongs to, we must be able to identify its kinematic structure. Like the human arm, the robot arm is made up of a series of links and joints. The joint is the part of the arm that connects two links and allows *relative movement* between the connecting rods. [3] In order to determine the kinematic structure, we need to recognize the joint types used and the "degree of freedom" the arm has.

Each robot has a base that is normally *secured* to the floor. However, the base can also be secured to a wall or ceiling and can also be incorporated into a *gantry*. All give the advantage of saving floor space and improve the robots reach, thus giving the *end effector* (which is a general term used to describe the tooling on the end of the robots arm) greater manipulative power. The first link of the robot arm is connected to the base and the last link is connected to the end effector. In general, the more joints in the robot arm the greater the *dexterity* it has. The typical structure of industrial robot consists of 4 major components: the manipulator, the end effector, the *power supply* and the control system, as shown in Fig. 5-2.

Words and Expressions

manipulator [məˈnɪpjuleɪtər] n. 操纵者；操纵器
campaign [kæmˈpeɪn] n. 运动；战役
rod [rɑd] n. 杆，拉杆

Unit 5
Industrial Robot

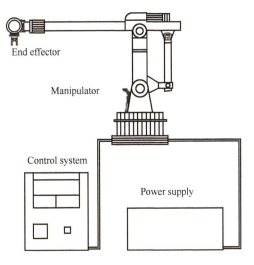

Fig. 5-2 Typical Structures of Robot

robotics [rəʊ'bɒtɪks] n. 机器人技术
secure [sɪ'kjʊə(r)] v. 获得安全，变得安全
gantry ['gæntri] n. 构台；桶架；台架
dexterity [dek'sterəti] n. 灵巧，熟练
relative movement 相对运动
end effector 终端执行器
power supply 电力供应

Special Difficulties

1. In contrast with the conventional view, robotics was not developed recently.
in contrast with 意为"相比之下"。
本句可译为：与常规观点相反，机器人学并非最近才发展起来的。

2. Today current research efforts focus on creating a "smart" robot that can "see" "hear" "touch" and consequently make decisions.
本句可译为：当前，对于机器人的研究集中在制造"智能型"机器人，集"视""听"和"触摸"等功能于一身，进而能做出决定。

3. The joint is the part of the arm that connects two links and allows relative movement between the connecting rods.
that 引导定语从句，对 the arm 进行说明。
本句可译为：关节是手臂中连接两个连杆的部件，它允许各连杆之间存在相对运动。

Learn and Practice

1. Mark the following statements with T (true) or F (false) according to the text.

1) In fact, the first robots were created in the early 1960's in America. (   )

2) In 1960 Cincinnati Millicron was the first mini-computer controlled robot. (   )

3) A robot always obeys human beings. (   )

4) In 1974 the IRB6 robot was introduced by the Swedish company ASEA. (   )

5) Like the human arm, the robot arm is made up of a series of links and joints. (   )

2. Choose the best choices according to the text.

1) A robot is a multi-functional manipulator designed to do things except (   ).

   A. move materials   B. move parts   C. self-study

2) The (   ) can connects two links and allows relative movement between the connecting rod.

   A. arm   B. joint   C. rod

3) The following components belong to the typical structure of industrial robot except (   ).

   A. the manipulator

   B. the end effector

   C. the power supply

   D. the arm

3. Translate the following phrases into English or Chinese.

1) 工业机器人。

2) 电力供应。

3) 传感技术。

4) kinematic structure.

5) degree of freedom.

Extensive Reading

Transfer Machine

The highest degree of automation with special purpose, multifunction machines is achieved by using *transfer machine*. Transfer machine is essentially a combination of individual workstations arranged in the required sequence, connected by work transfer devices, and integrated with interlocked controls (Fig. 5-3). [1] Workpieces are automatically transferred between the stations, which are equipped with horizontal, vertical, or *angular* units to perform machining, *gaging*, *workpiece repositioning*, assembling, washing, or other operations. [2] The two major classes of transfer machines are rotary and in-line types.

An important advantage of transfer machines is that they permit the maximum number of operations to be performed simultaneously. There is relatively no limitation on the number of workpiece surfaces or planes that can be machined, since devices can be interposed in transfer machines at practically any point for inverting, rotating, or orienting the workpiece, so as to complete the machining operations. Work repositioning also *minimizes* the need for angular machining heads and allows

Unit 5
Industrial Robot

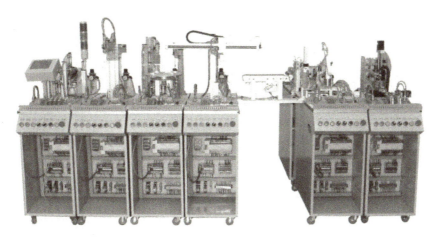

Fig. 5-3 *Modular production system*

operations to be performed in optimum time. Complete processing from rough *casting* or *forging* to finished parts is often possible.

All types of machining operations, such as drilling, *tapping*, *reaming*, boring, and milling, are economically combined on transfer machines (Fig. 5-4). Lathe type operations such as turning and facing are also being performed on in line transfer machines, with the workpiece being rotated in selected machining stations. Turning operations are performed in lathe type segments in which *toolholders* are fed on slides mounted on tunnel type bridge units. workpiece are located on centers and rotated by chucks at each turning station. Turning stations with CNC are available for use on in line transfer machine. The CNC units allow the machine cycles to be easily altered to accommodate changes in workpiece design and can also be used for automatic tool adjustments.

Fig. 5-4 *Transfer machine in industry*

Maximum *production* economy on transfer lines is often achieved by assembling parts to the workpieces during their movement through the machined. Such items as bushings, seals, Welch plugs, and heat tubes can be assembled and then machined or tested during the transfer machining sequence. Automatic nut *torquing* following the application of part subassemblies can also be carried out.

Gun drilling or reaming on transfer machines is an ideal application provided that proper machining units are employed and good bushing practices are followed. [3] *Contour* boring and turning of spherical seats and other surfaces can be done with tracer controlled single point inserts, thus eliminating the need for costly special form tools. In process gaging of reamed or bored holes and automatic tool setting are done on transfer machines to maintain close tolerances.

Transfer machines have long been used in the automotive industry for high production rates with a minimum of manual part handling. In addition to decreasing labor requirements, such machines ensure consistently uniform, high quality parts at lower cost. [4] They are no longer confined just to rough machining and now often eliminate the need for subsequent operations such as grinding and honing. [5]

Words and Expressions

angular [ˈæŋgjələr] adj. 有角的，用角测量的
gage [geidʒ] vt. 计量，度量
minimize [ˈmɪnɪmaɪz] vt. 把……减至最低数量（程度）
cast [kɑːst] vt. 铸造
forge [fɔːdʒ] vt. 锻造；伪造
tap [tæp] v. 切割
ream [riːm] v. 扩孔
toolholder [ˈtuːlhəʊldər] n. 刀柄
production [prəˈdʌkʃn] n. 生产，制作
torque [tɔːk] n. （尤指机器的）扭转力，转（力）矩
contour [ˈkɒntʊər] n. 外形，轮廓
transfer machine 自动生产线
modular production system 模块化生产系统
workpiece reposition 工件再定位

Special Difficulties

1. Transfer machine is essentially a combination of individual workstations arranged in the required sequence, connected by work transfer devices, and integrated with interlocked controls.

本句可译为：自动生产线实质上是各个独立工作站的组合体，这些工作站按需求顺序排列，由工件传送装置连接，并且通过连锁控制集成。

2. Workplaces are automatically transferred between the stations, which are equipped with horizontal, vertical, or angular units to perform machining, gaging, workplace repositioning, assembling, washing, or other operations.

Unit 5
Industrial Robot

be equipped with... 意为"安装有……, 装配有……"。

本句可译为：工件在工位间被自动传送，每个工位都装配有用于加工、测量、工件再定位、组装、清洗及其他操作的卧式、立式及倾斜式的设备。

3. Gun drilling or reaming on transfer machines is an ideal application provided that proper machining units are employed and good bushing practices are followed.

provided that 意为"假设，如果"，引导宾语从句。

本句可译为：如果能使用合适的机加工装置且随后进行良好的操作，在自动生产线上进行深钻或铰孔是一项理想的应用。

4. In addition to decreasing labor requirements, such machines ensure consistently uniform, high quality parts at lower cost.

in addition to 意为"除了"。

本句可译为：除了减少劳动力需求外，自动生产线可始终保证以更低的成本生产符合标准的优质零件。

5. They are no longer confined just to rough machining and now often eliminate the need for subsequent operations such as grinding and honing.

no longer 意为"不再"。

本句可译为：它们不再局限于粗加工，现在通常不再需要诸如研磨和珩磨这样的后续工序。

Learn and Practice

1. Mark the following statements with T (true) or F (false) according to the text.

1) According to the text, transfer machines are essentially a combination of individual workstations arranged in the required sequence, connected by work transfer devices, and integrated with interlocked controls. （　）

2) An important advantage of transfer machines is that they permit the maximum number of operations to be performed interrupted. （　）

3) According to the text, the two major classes of transfer machines are rotary and in-line types. （　）

4) All types of machining operations, such as drilling, tapping, reaming, boring, and milling, are economically combined on transfer machines. （　）

5) Minimum production economy on transfer lines is often achieved by assembling parts to the workplaces during their movement through the machine. （　）

2. Translate the following phrases into English or Chinese.

1) 自动生产线。

2) 成品工件。

3) contour boring.

4) labor requirements.

Knowledge Link

科技英语的词汇构成和翻译 1

一、科技英语词汇概述

科技英语的词汇是用来记录和表述各种现象、过程、特性、关系、状态等不同内容的名称，亦称术语（term）。它在通用的词汇中虽占少数，但举足轻重，往往是段落或文章论述内容的中心词语，如 microprocessor（微处理器），welding（焊接），NC（数控）等技术词汇。科技英语词汇的主要特征如下：

（1）涌现性　新学科的诞生、新化合物的合成、新物种的发现等，使科技术语随着技术的发展而不断出现。

（2）单义性　对于某一特定专业或其分支，科技术语词义狭窄，形态单一，定义时尽可能避免"同形异义"或"同义异形"现象。

（3）中性　术语只有概念意义，没有任何附加色彩。如 cat 转译为"吊锚"、dog 转译为"卡爪器"后，就失去了原来猫和狗的形象，及人们在用词方式上可能反映的好恶。

（4）国际性　科技英语和其他印欧语系中的部分技术词和半技术词汇来源于拉丁语和希腊语，且词汇的专业性越强，在印欧语系中"同形同义"词越多，因而国际性也越强。另外，新创造的科技词语只要在某一国使用和流行，英语国家和其他国家就可按照发音对应规律和拼写体系将其转写过来，成为自己的词汇，例如，transistor（晶体管）一词在英语、德语和法语中是完全一样的。

二、科技英语词汇构成

科技英语随着社会和技术的进步而发展，词汇的数量由少增多，词汇的含义由简单变复杂，符合语言发展的一般规律。科技英语词汇的来源及构成主要有以下四个方面：

1. 英语中的普通词汇，但在科技英语中被赋予了新的词义

Work is the transfer of energy expressed as the product of a force and the distance through. Which its point of application moves in the direction of the force.

在这句话中，"work" "energy" "product" "force" 都是从普通词汇中"借"来的物理学术语："work" 的意思不是"工作"，而是"功"；"energy" 的意思不是"活力"，而是"能"；"product" 的意思不是"产品"，而是"乘积"；"force" 的意思不是"力量"，而是"力"。

2. 来自于希腊语或者拉丁语

例如，therm（热，希腊语），thesis（论文，希腊语），parameter（参数，拉丁语），radius（半径，拉丁语）等词都是这种类型。这些来源于希腊语和拉丁语的词的复数形式有些仍按原来的形式，如 thesis 的复数形式是 theses，stratus 的复数形式是 strati。但由于在英语里使用时间较长，不少词除保留了原来的复数形式外，又采用了英语的复数形式，例如：

formula（公式，拉丁语）的复数形式可以是 formulae，也可以是 formulas；stratum（层，拉丁语）的复数形式可以是 strata，也可以是 stratums。特别有意思的是，datum（数据，拉丁语）的复数形式是 data，但是由于这个词经常用复数，所以有人会误把它当成单数，又造出了一个所谓的复数形式 datas。

3. 新出现的词汇

随着科学技术的不断发展，许多新的事物和复杂的概念也不断产生，原有的词汇已不够用，人们便创造了一些新词来表示新的事物和概念。按照一定的语言规律创造新词的方法，叫作构词法（word-formation or word-building）。英语中的构词法主要有以下几种：

（1）转化　通过词类转化构成新词。

英语中名词、形容词、副词、介词可以转化成动词，动词、形容词、副词、介词可以转化成名词。但最活跃的是名词转化成动词和动词转化成名词，例如，名词 island（小岛）转化成动词 island（隔离），动词 coordinate（协调）转化成名词 coordinate（坐标）。

（2）合成　由两个独立的词合成为一个新词。

例如：air + craft→aircraft（飞机），air + port→airport（机场），metal + work→metalwork（金属制品），power + plant→powerplant（发电站）。有的合成词的两个成分之间要有连字符，例如：cast-iron（铸铁），conveyer-belt（传送带），machine-made（机制的）。

英语中有很多专业术语由两个或更多的词组成，叫作复合术语。它们的构成成分虽然看起来是独立的，但实际上是合起来构成一个完整的概念，因此应该把它们看成是一个术语。例如：liquid crystal（液晶），computer language（计算机语言），machine building（机器制造），linear measure（长度单位），civil engineering（土木工程）。

（3）派生　这种方法也叫缀合。派生词是由词根加上前缀或后缀构成的。

添加前缀构成新词，只改变词义，不改变词性。例如：decontrol（取消控制，动词）→de + control（control 是动词），ultrasonic（超声的，形容词）→ultra + sonic（sonic 是形容词），subsystem（分系统，名词）→sub + system（system 是名词）。

有些加前缀的派生词在前缀和词根之间有连字符，例如：hydro-electric（水力发电），extra-terrestrial（外行星际的）。英语的前缀是有固定意义的，记住一些常用前缀对于记忆生词和猜测词义有帮助。

英语后缀的作用和前缀有所不同，它们主要用来改变词性，从一个词的后缀可识别它的词性，这是后缀的语法意义。后缀的词汇意义往往并不明显，因为有时加后缀可能不改变词义，无法创造出我们需要的新词。

（4）缩略　把词省略或者简化，构成新词。现在的趋势是缩略词的数目不断增多，使用范围不断扩大。常见的缩略形式有以下四种：

1）缩略词的首部：telephone = phone，university = varsity，helicopter = copter，etc.

2）缩略词的尾部：advertisement = ad，debutante = deb，modern = mod，professional = pro，exposition = expo，memorandum = memo，etc.

3）缩略词的首尾：influenza = flu，detective = tec，refrigerator = fridge，etc.

4）首字母缩略：VOA = Voice of America，IOC = International Olympic Committee，WTO = World Trade Organization，CAD = Computer Aided Design，NATO = North Atlantic Treaty Organization，CAE = Computer Aided Engineering，CAM = Computer Aided Manufacture，etc.

4. 约定的专业术语

在科技英语中大量出现专业术语，专业术语语义较固定，翻译也是固定的表达方式。例如：TCP/IP（Transmission Control Protocol/Internet Protocol，传输控制协议/互联网协议），relay（继电器），interface（通信接口），point-to-point（点对点），topology（拓扑学），HTML（Hypertext Markup Language，超文本标识语言），HTTP（Hypertext Transfer Protocol，超文本传输协议），LAN（Local Area Network，局域网），Modem（调制解调器），aerodynamics（空气动力学），cybernetics（控制论），precision（精度），vector（向量或矢量），matrix（矩阵），prototype（原型，样机），state-of-the-art（工艺水平），detect（探测，检测），fault（故障），hydraulic system（液压系统），NC（Numerical Control，数控），milling（铣），feed（进给量），machine tool（机床），feedback（反馈），hardness（硬度），coolant（冷却液），roughness（粗糙度），sensor（传感器），integration（集成，综合），pulse train（脉冲序列），patent（专利），poly-phase（多相的），transformer（变压器），integrate circuit（集成电路）。

值得一提的是，科技文章中有些词汇并非专业术语，除在科技文章中使用外也广泛应用在政治、经济、法律、语言等社会科学的文章当中。例如：accordance（按照），imply（隐含），acknowledge（承认），inclusion（包括），alter（改变），incur（招致），alternative（交替的），indicate（指示），amend（修正），induce（导致），application（应用），initial（初始的），appropriate（恰当的），modification（修改），attain（达到），nevertheless（然而），circumstance（情况），nonetheless（然而），compensation（补偿），obtain（获得），confirm（证实），occur（发生），consequence（后果），omission（省略），considering（鉴于），providing（假设），consist（组成），reduce（减少），constitute（构成），replacement（代替），consume（消耗），specifically（具体地说），deduce（推理），distinction（差别），valid（有效的），function（功能），verify（验证），illustrate（说明）等。

Dialogue Exercise

Background：Li Lei and Han Meimei are now visiting an industrial robot show in a university. A young girl named Lucy is in charge of showing the visitors around.

Lucy：Ladies and gentlemen, welcome to our industrial robot show. I'm Lucy. My job is to show you around. You can ask me about anything you do not understand.

Han Meimei：How interesting to see so many kinds of robots here. Excuse me, Lucy, but I wonder why these robots do not look like nor behave like human beings?

Lucy：Ah, I see. The point here is that what we show you are not other kinds of robots but industrial ones. We needn't make them look like us humans.

Han Meimei：Well, I understand. Now could you tell us what these industrial robots can do for us?

Lucy：Sure. These robots can do many things for us. They can help us, for example, to handle materials, spray paint, rescue people from the fire, and even help to explore deep

Unit 5
Industrial Robot

oceans and outer-space.

Li Lei: Uh, how great robots are! But what makes robots able to work for us so marvelously?

Lucy: It's nothing else but the computer programs. Robots are a programmable mechanical manipulator. Being programmed with human instructions, they can move along several directions and do factory work usually done by human beings.

Han Meimei: Oh, that's wonderful. But what's the difference between robot and my handy calculator or my home washing machine, now that all of them belong to some reprogrammable mechanical manipulator.

Lucy: Okay, it's a clever question. But I think, here you mix a robot with an ordinary washing machine because you forget a robot can do a stand-alone operation, that is, rather independent jobs, while a washing machine can't. So when we talk about industrial robot, we mean a machine or device, which is something reprogrammable with its own end effector, which can perform a factory duty in a stand-alone manner.

Li Lei: Oh, I guess, my automatic watch and my home washing machine cannot play chess with me while a robot can.

Lucy: That's right. You're so smart.

Famous Quotes

Science surpasses the old miracles of mythology.
科学远远地超过了那些神话中的古老奇迹。

Unit 6　CIMS

知识目标:

1. 了解计算机集成制造系统的发展。
2. 掌握计算机集成制造系统的工作原理和相关专业术语。
3. 掌握科技英语的词性变化和词义变化。

能力目标:

1. 能对计算机集成制造系统的专业术语进行中英互译。
2. 能对计算机集成制造系统相关英文资料进行阅读和翻译。
3. 能正确翻译科技英语文章中的多义词。

Reading Material

INTRODUCTION: CIMS is an advanced technology that produced with the development of the computer. Relying on computer technology, it integrated various isolated automation subsystems, dispersed in the product design and manufacturing process, forming an integrated and intelligent manufacturing system that is suitable for multi-variety and small-batch production.

CIMS is the abbreviation of Computer Integrated Manufacturing System. It is produced with the development of CAD and CAM. CIMS describes a new *approach* to manufacturing, management, and corporate operation. [1] Although CIM systems may include many advanced manufacturing technologies such as robotics, Computer Numerical Control (CNC), Computer-aided Design (CAD), Computer-aided Manufacturing (CAM), Computer-aided Engineering (CAE), and Just-in-Time Production, it goes beyond these technologies (Fig. 6-1).

CIMS manufacture is a fully automated system and computer software covering all the three major areas of CIMS: factory automation, production and process design, and manufacturing planning and control.

CIMS includes the software and automation systems needed to perform the entire CIMS process

Fig. 6-1   CIMS

(Fig. 6-2). It includes product design, system programming, *estimation* of production costs, actual manufacturing of products, order entry, inventory tracking, and analyses of the actual manufacturing costs.

CIMS is a real kind of *flexible* manufacturing system and can manufacture a wide variety of parts or assemblies in small-batch quantities with no physical changes to the system. This is possible because using software the programmer can easily program the *conveyor*'s material routings, robot programs and CNC machine programs, the automatic storage and *retrieval* system. Under the control of programs, the automatic storage and retrieval system can provide many different parts. [2] And finally the flexible conveyor uses carrier pallets and handles many different types of materials by using *interchangeable* tools.

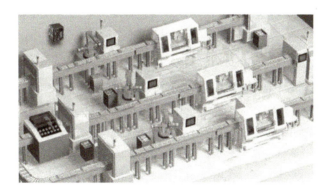

Fig. 6-2   The application of CIMS

Words and Expressions

approach [əˈprəʊtʃ] n. 方法；接近
estimation [ˌestɪˈmeɪʃn] n. 估算；评估
flexible [ˈfleksəbl] adj. 灵活的；柔性的

conveyor [kən'veɪər] n. 传输；输送机

retrieval [rɪ'triːvl] n. 恢复；取回

interchangeable [ˌɪntə'tʃeɪndʒəbl] adj. 可交换的；可交替的

Special Difficulties

1. CIMS describes a new approach to manufacturing, management, and corporate operation.

本句可译为：CIMS描述了一种制造、管理和企业运营的新方法。

2. Under the control of programs, the automatic storage and retrieval system can provide many different parts.

under the control of... 意为"在……控制下，受……的控制"。

本句可译为：在程序的控制下，自动存储和检索系统可提供许多不同的零件。

Learn and Practice

1. Mark the following statements with T (true) or F (false) according to the text.

1) CIMS is the abbreviation of Computer Integrated Manufacturing System. (   )

2) CIMS manufacture is a fully automated system and computer software covering all the four major areas of CIMS. (   )

3) CIMS includes the software and automation systems needed to perform the entire CIMS process. (   )

2. Translate the following phrases into Chinese or English.

1) Agile Manufacturing.

2) Virtual Manufacturing.

3) 计算机辅助设计。

4) 计算机辅助工程。

3. Translate the following sentences into Chinese.

CIMS is a real kind of flexible manufacturing system and can manufacture a wide variety of parts or assemblies in small-batch quantities with no physical changes to the system. This is possible because using software the programmer can easily program the conveyor's material routings, robot programs and CNC machine programs, the automatic storage and retrieval system.

Extensive Reading

CNC Machine Tools

NC is being used in all types of machine tools, from the simplest to the most complex. NC machine tools were developed on the basis of general-purpose machine tools. Various types of NC machine tools originated from the same types of general-purpose machine tools. Since NC machine tools have been developed for over forty years, they have many specifications and types, and their structures and functions have their own features. In order to understand NC machine tools, we introduce several main CNC machine tools simply.

1. NC *Turning Machine*

The Fig. 6-3 shows the appearance of a NC turning machine. It is one of the most productive machine tools. The body of NC turning machine includes a *spindle*, an *apron* and a tool post. NC system consists of CRT display, control panel and heavy-current control system.

NC turning machine has the function of simultaneous two-axis movement. Z-axis is parallel to the spindle and X-axis is perpendicular to the spindle in the horizontal plane. The Z-axis controls the *carriage* travel toward or away from the headstock. The X- axis controls the *cross motion* of the cutting tool. And with the addition of C-axis in the newest turning and *milling machining center*, the machine is used for workpieces indexing and milling by fixing milling cutters in the carriage.

2. NC Milling Machine

The NC milling machine (Fig. 6-4) has always been one of the most versatile machines used in industry. Operations such as milling, *contouring*, *gear cutting*, *drilling*, *boring*, and *reaming* are only a few of the many operations. NC milling machine is suitable for processing 3D complex *curved surface*, finding wide application in automobile, *aerospace*, die equipment. In the world the first NC machine tool came from NC milling machine. But with the development of times, NC milling machine tends to machining center. At present, because of low price, convenient and flexible operation, short time for working preparation, NC milling machine is still widely in use, classified as NC *vertical milling machine*, *horizontal milling machine* and NC copying milling machine.

Fig. 6-3 NC turning machine

Fig. 6-4 NC milling machine

3. Machining center

Machining center is a product which is formed when NC machine tools are developed to a certain stage (Fig. 6-5). So far, machining center isn't clearly defined. NC boring and milling machines equipped with automatic tool changers are generally considered as machining centers. [1] Actually, machining center is summed up in the phrase "with automatic tool changer, capable of more processes machining." Machining center can perform milling, boring, drilling, reaming and tapping, which eliminates the need for a number of individual machine tools. Thus, it reduces *capital and labor requirements*.

Machining centers can be divided into the vertical machining center and the horizontal machining centers. The spindle of vertical machining center is perpendicular and the spindle of horizontal machining center is horizontal. Vertical machining centers continue to be widely accepted and used, primarily for flat parts and where three axis machining is required on a single part face such as mould and die work. Horizontal machining centers are also widely accepted and used, particularly with large, boxy, heavy parts because they lend themselves to easy and accessible *pallet* shuttle transfer when used in a cell or FMS application.

A workpiece can be put on a rotating table or exchanging pallet by using a fixture. Polyhedrons can be produced by the table rotating, and the exchanging of the pallet can change the machining workpieces for improving the machining efficiency.

4. NC Drilling Machine

The Fig. 6-6 shows an NC drilling machine. It can be divided into vertical NC drilling machine and horizontal NC drilling machine. NC drilling machine can perform drilling and tapping, also simple milling. Its *tool magazine* can store various tools.

Fig. 6-5　Machining center

Fig. 6-6　NC drilling machine

5. NC *Grinding Machine*

NC grinding machines (Fig. 6-7) are used to machine high hardness and precision machining surfaces. They are classified as NC surface grinders, NC internal grinders and NC profile grinders.[2] With the development of automatic grinding wheel compensation, correction and grinding fixed cycle technology, the function of NC grinding machine is stronger and stronger.

6. NC *Electrical Discharge Forming Machine*

Every machinist knows that on standard machine tools electrical energy is converted into motion by an electric motor. Nowadays, it has been discovered that electrical energy can be directly employed in metal removal. The NC electrical discharge forming machine (Fig. 6-8) is a special machining method, utilizing two different polarities of electrode to generate discharge in the insulator for removing material and finishing machining. It has the special advantage in machining complex die and difficult-to-machine materials.

Fig. 6-7 NC grinding machine

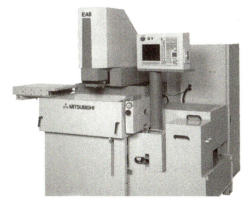

Fig. 6-8 NC electrical discharge forming machine

7. NC *Wire-cut Machine Tool*

The working principle of NC wire-cut machine tools (Fig. 6-9) is the same as the principle of electrical discharge forming machine tools. Its electrode is electrode wire and it generally uses deionized water as the processing liquid. [3]

Fig. 6-9 NC wire-cut machine

Words and Expressions

turning ['tɜːnɪŋ] n. 车削
spindle ['spɪndl] n. 主轴
apron ['eɪprən] n. 溜板
carriage ['kærɪdʒ] n. 刀架
milling ['mɪlɪŋ] n. 铣削
contouring [kən'tʊərɪŋ] n. 轮廓，造型

drilling ['drɪlɪŋ] n. 钻削
boring ['bɔːrɪŋ] n. 镗削
reaming ['riːmɪŋ] n. 铰削
aerospace ['eərəʊspeɪs] n. 航空航天工业
pallet ['pælət] n. 托盘，托板；平台
turning machine 车床
cross motion 横向运动
machining center 加工中心
gear cutting 齿轮加工
curved surface 曲面
vertical milling machine 立式铣床
horizontal milling machine 卧式铣床
capital and labor equipments 资金和劳动力需求
tool magazine 刀库
grinding machine 磨床
electrical discharge forming machine 电火花成型机床
wire-cut machine tool 线切割机床

Special Difficulties

1. NC boring and milling machines equipped with automatic tool changers are generally considered as machining centers.

be equipped with 意为"配备有……"，equipped with automatic tool changers 是过去分词短语做后置定语，修饰 NC boring and milling machines。

本句可译为：装配有自动换刀装置的数控镗铣床，通常被认为是加工中心。

2. They are classified as NC surface grinders, NC internal grinders and NC profile grinders.

be classified as 意为"被分为……"。

本句可译为：它们（数控磨床）可分为数控平面磨床、数控内圆磨床和数控仿形磨床。

3. Its electrode is electrode wire and it generally uses deionized water as the processing liquid.

deionized water 意为"去离子水"。

本句可译为：其电极是电极丝，加工液一般采用去离子水。

Learn and Practice

1. Mark the following statements with T (true) or F (false) according to the text.

1) In NC turning machine, X-axis is parallel to the spindle and Z-axis is perpendicular to the spindle in the horizontal plane. (   )

2) The NC milling machine has always been one of the most versatile machines used in industry. (   )

3) The spindle of vertical machining center is horizontal and the spindle of horizontal machining center is perpendicular. (　　)

4) Electrical discharge forming machine has the special advantage in machining simple die and difficult-to-machine materials. (　　)

5) NC milling machines are used for machining 3D complex curved surface. (　　)

6) The working principle of NC wire-cut machine tools is not the same as the principle of electrical discharge forming machine tools. (　　)

2. Choose the best choices according to the text.

1) NC turning machine has the function of simultaneous (　　)-axis movement.

A. two　　　　　　　B. four　　　　　　　C. five

2) NC milling machine has always been one of the most versatile (　　) used in industry.

A. machining　　　　B. machines　　　　　C. machine

3) On standard machine tools electrical energy (　　) into motion by an electric motor.

A. convert　　　　　B. is converted　　　　C. converts

4) In the following aspects, all are correct except (　　).

A. In NC turning machine, Z-axis is perpendicular to the spindle.

B. Horizontal machining centers are used for machining large, boxy, heavy parts.

C. NC electrical discharge forming machines have the special advantage in machining complex die and difficult-to-machine materials.

5) NC milling machine is still widely in use, (　　) as NC vertical milling machine, horizontal milling machine and NC copying milling machine.

A. classify　　　　　B. to classifying　　　C. classified

3. Translate the following phrases into English or Chinese.

1) 数控车床。

2) 数控铣床。

3) 加工中心。

4) electrical discharge forming machine.

5) grinding machine.

6) tool magazine.

Knowledge Link

科技英语的词汇构成和翻译 2

一、词义的引申

英、汉两种语言在表达方式上差别很大。翻译时，有些词或词组无法直接搬用辞典中的释义，若勉强按辞典中的释义逐词死译，会使译文生硬晦涩，很难读懂，甚至会造成误解。

所以，要在弄清原文词义的基础上，根据上下文的逻辑关系和汉语的搭配习惯，对词义加以引申。若遇到有关专业方面的内容，必须选用专业方面的常用术语。引申后的词义虽然与辞典中的释义稍有不同，但却能更确切地表达原文意义。示例如下：

1) There is no physical contact between tool and work piece.

直译：在工具和工件之间没有有形的接触。

引申：工具和工件不直接接触。

2) Public opinion is demanding more and more urgently that something must be done about noise.

直译：公众舆论越来越强烈地要求为消除噪声做某些事情。

引申：公众舆论越来越强烈地要求管一管噪声问题。

3) There is a wide area of performance duplication between numerical control machines and automatics.

直译：在数控机床和自动化机床之间，有一个性能重复的广阔地带。

引申：数控机床和自动化机床有很多相同的性能。

4) High-speed grinding does not know this disadvantage.

直译：高速磨床不知道这个缺点。

引申：高速磨床不存在这个缺点。

二、词性的变化

原文中有些词在译文中需要转换词性，才能使译文通顺自然。词性转译大体有以下几种情况：

1. 英语动词、形容词、副词译成汉语名词

1) Gases differ from solids in that the former has greater compressibility than the latter.

本句可译为：气体和固体的区别在于气体的可压缩性比固体大。

2) The instrument is characterized by its compactness, and portability.

本句可译为：这个仪器的特点是结构紧凑、携带方便。

(3) The cutting tools must be strong, tough, hard and wear resistant.

本句可译为：刀具必须有足够的强度、韧性、硬度，而且耐磨。

2. 英语名词、介词、形容词、副词译成汉语动词

1) The application of electronic computers makes for a tremendous rise in labor productivity.

本句可译为：使用电子计算机可以大大提高劳动生产率。

2) In any machine input work equals output work plus work done against friction.

本句可译为：任何机器的输入功，都等于输出功加上克服摩擦所做的功。

3) Scientists are confident that all matter is indestructible.

本句可译为：科学家们深信一切物质是不灭的。

3. 英语的名词、副词和动词译成汉语形容词

1) The maiden voyage of the newly-built steamship was a success.

本句可译为：那艘新造轮船的处女航是成功的。

2) It is a fact that no structural material is perfectly elastic.

本句可译为：事实上没有一种结构材料是十全十美的弹性体。

3）They said that such knowledge is needed before they develop a successful early warning system for earthquakes.

本句可译为：他们说，这种知识对他们发明一种有效的地震早期警报系统是必要的。

4. 英语形容词、名词译成汉语副词

1）A continuous increase in the temperature of the gas confined in a container will lead to a continuous increase in the internal pressure within the gas.

本句可译为：不断提高密封容器内气体的温度，会使气体的内压力不断增大。

2）It is our great pleasure to note that China will host the Winter Olympic Games.

本句可译为：我们很高兴地注意到，中国即将主办冬奥会。

三、词义的差异

英语单词绝大多数为多义词，但科技英语文章的翻译要求严谨准确，因此翻译时首先要选择最确切的词义。只有这样，才能使译文正确。选择词义通常从以下几个方面考虑：

1. 根据词类的差异确定词义

同一个单词，词类不同，其词义也不同。例如，like 在下面三个句子中分属三种不同词类，弄错了词类，理解上就错了，译文也错了。

1）Like charges repel, while unlike charges attract.

本句可译为：同性电荷相斥，异性电荷相吸。like 是形容词，意为"同样的"。

2）Things like air, water or metals are matter.

本句可译为：像空气、水或金属之类的东西都是物质。like 是介词，意为"像，如"。

3）I hope I can use the computer like you do.

本句可译为：我希望我能像你一样使用计算机。like 是连接词，意为"如同，像……一样"。

2. 同一个词即使词类相同，词义也可能有差别

在下面三个句子中，light 均为形容词，但词义不相同。

1）The instrument is very light.

本句可译为：这台仪器很轻。light 意为"轻的"。

2）The cover of the meter is light blue.

本句可译为：这个仪表盖是浅蓝色的。light 意为"浅的"。

3）The lamp is very light.

本句可译为：这盏灯是很明亮的。light 意为"明亮的"。

3. 根据名词的单/复数确定词义

英语中有些名词的单数和复数所表示的词义完全不同。例如：

名　　词	单 数 词 义	复 数 词 义
facility（ies）	简易，灵巧	设施，工具
charge（s）	负荷，电荷	费用
main（s）	主线，干线	电源
spirit（s）	精神	酒精
work（s）	功，工作	著作，工厂，工程

四、词汇的搭配

英语和汉语的词汇各有其搭配规律,在翻译时如果不分具体场合,机械地复制原文里的词汇搭配,孤立地译出每个词的字面意义,往往会破坏译文语言的规范,得不出正确的译文。因此,有时必须根据原文的上下文逻辑关系和译文语言习惯,适当地调整原文的字面意义,换用另一些词来表达。

1. 动词与名词的搭配

以动词"move"为例,其基本意思是"运动",但在不同的句子里与不同的名词搭配,译法也就不同了:

1) The earth moves round the sun.

本句可译为:地球<u>绕太阳旋转</u>。

2) Heat moves from a hotter to colder body.

本句可译为:热量从温度较高的物体<u>传到温度较低的物体上</u>。

3) Move M to other work.

本句可译为:<u>调</u> M 去做别的工作。

2. 副词与动词的搭配

以副词"successfully"为例:

1) Spot-welding has already been used successfully in welding fuel tanks.

本句可译为:点焊已经<u>成功地用来焊接油箱</u>。

2) The task was finished successfully.

本句可译为:<u>圆满完成</u>了任务。

3) Our production plan has been successfully carried out.

本句可译为:我们的生产计划已<u>顺利执行</u>了。

3. 形容词与名词的搭配

以形容词"thick"为例:

1) This is a 8 inches thick book.

本句可译为:这是本 8 英寸<u>厚的书</u>。

2) He drew a thick line on the paper.

本句可译为:他在纸上画了一条<u>粗线</u>。

3) Thick liquids pour much slowly than thin liquids.

本句可译为:<u>稠的液体</u>比稀的液体倒得慢。

4. 动词与介词、动词与副词的搭配

科技英语中大量出现动词与介词,及动词与副词的搭配使用,这些搭配的意思常常是固定的,需要牢记。

常见的搭配有:set up(建立), differ from(不同,区别于), be characterized by(以……为特征), find out(发现,找到), work out(计算出,设计出,解决), figure out(计算出,解决), apply for(请求,申请), be applied to(应用到), be divided/separated into(分成,分为), used up(用完,耗尽), run out(用完,耗尽), give off(放出光、热), remove from(从……取下,脱下), study/research on(研究……), be based on(基于,以……为基

础）等。

Dialogue Exercise

Background: Li Lei and Han Meimei are now paying a visit to the tool room of the school factory. Mr. Wang, an old master comes over and shows them around.

Wang: Hello, everyone, welcome to this tool room. The tools, devices and all kinds of simple machine that you see are free for you to touch. While visiting, you can ask me any question. Now, come along with me and let me show you around.

Li Lei: Mr. Wang, look at the saws. Why are the two saws different both in shape and size?

Wang: Yeah. You know, the bigger one near you is called a handsaw. It is usually used by a carpenter for cutting wood. The other one over there is named a hacksaw. It is used for sawing metal pipes or ceramic titles. A hacksaw is different from a handsaw in that it is adjustable.

Han Meimei: Mr. Wang, how interesting to see all those wrenches! There are so many kinds and they're quite different in size and use.

Wang: You're right. They are various both in size and look because they are used to do different things. You see, this is a 6-inch wrench. It is used for tightening furniture bolts, toilet bolts and the like. It is unadjustable, you have to use it when you repair your bikes. That one over there is a 10-inch wrench. It is adjustable. You can use an adjustable wrench to repair a car.

Li Lei: Mr. Wang, I like the drills hanging over there. It seems to me that a drill, especially an electric one, is so much different from a screwdriver. It can move so fast like a machine.

Wang: That sounds reasonable. You can call a drill a simple machine for it can produce a movement and work on something else.

Li Lei: Excuse me, Mr. Wang, may I ask you which of these tools we will use most often?

Wang: It's up to the work you're going to do. Actually, a mechanic must be good at using all kinds of tools here you see.

Li Lei & Han Meimei: Thank you very much, we've learned quite a lot here, indeed.

Wang: You're welcome.

Famous Quotes

Keep trying no matter how hard it seems, it will get easier.

坚持不懈，难也变易。

Unit 7　Mechatronics Technology

知识目标:

1. 了解机电一体化技术的发展及相关专业术语。
2. 掌握机电一体化技术的应用领域。
3. 掌握科技英语的常见表达方式。

能力目标:

1. 能对机电一体化技术的专业术语进行中英互译。
2. 能对机电一体化技术相关英文资料进行阅读和翻译。
3. 能准确理解科技英语的时态、语态特点。

Reading Material

INTRODUCTION: *Mechatronics* has been adapted as the *synergetic* integration of three disciplines: mechanics, control and electronics. This paper states that the history of mechatronics technology development and the application of mechatronics in industry.

A Japanese engineer from Yasukawa Electric Company coined the term "mechatronics" in 1969 to reflect the merging of mechanical and electrical engineering disciplines.[1] Until the early 1980s, mechatronics meant a mechanism that was electrified. In the mid-1980s, mechatronics came to mean engineering that was the *boundary* between mechanics and electronics. Today, the term *encompasses* a large array of technologies, many of which have become well-known in their own right. Each technology still has the basic element of the merging of mechanics and electronics but now many also involve much more, particularly software and information technology. For example, many early robots resulting from mechanical and electrical systems became central to mechatronics (Fig. 7-1).

As shown in Fig. 7-2, mechatronics is the *interdisciplinary fusion* of mechanics, electronics and information technology. The objective is for engineer to complete development, which is why it is currently so popular with industry. Mechatronics development so far has also become a subject having

Unit 7
Mechatronics Technology

Fig. 7-1 The mechatronics devices

its system of new disciplines, along with the science and technology development, but also not only will be with new contents. Its basic characteristics can be summarized as mechatronics from the *viewpoint* of system, comprehensive use of mechanical technology, *microelectronics technology*, automatic control technology, computer technology and information technology, sensing measurement and control technology, *power electronic technology*, *interface technology*, information transform technique and *software programming technology*. According to the system function target and optimize organizational goals, rational configurations of each functional units, with the layout in multi-function, high quality, high reliability, low energy *consumption* on the meaning of achieving specific function value, make the whole system optimization. This produces function system, becoming a mechatronics system or the mechatronics product.

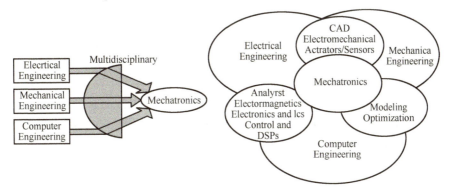

Fig. 7-2 The interdisciplinary nature of Mechatronics

In the *domain* of factory automation, mechatronics has far reaching effects in manufacturing and will gain even more importance in future. Major *constituents* of factory automation include CNC machines, robots, automation system, and CIMS. Basically these advanced manufacturing solutions consist of mechatronic systems. Low-volume, more variety, higher levels of flexibility, reduced lead time in manufacture, and automation in manufacturing and assembly are likely to be the future needs of customers. Evidently, the design and manufacture of future products will involve a combination of precision mechanical and electronic systems, and mechatronics will form the core of all activities in products and production technology.[2]

Words and Expressions

mechatronics [ˌmekəˈtrɔniks] n. 机电一体化
synergetic [ˌsinəˈdʒetik] adj. 协同的；协作的
boundary [ˈbaundəri] n. 分界线；范围
encompass [enˈkʌmpəs] vt. 围绕，包围；完成
interdisciplinary [ˌintə(ː)ˈdisiplinəri] adj. 跨学科的；各学科间的
fusion [ˈfjuːʒən] n. 溶合；溶解；（物）聚合物
viewpoint [ˈvjuːpɔint] n. 观点，意见；角度
domain [dəuˈmein] n. 范围；领域
constituent [kənˈstitjuənt] adj. 构成的；有选举权的
consumption [kənˈsʌmpʃən] n. 消费，耗尽
microelectronics technology 微电子技术
power electronic technology 电力电子技术
interface technology 接口技术
software programming technology 软件编程技术

Special Difficulties

1. A Japanese engineer from Yasukawa Electric Company coined the term "mechatronics" in 1969 to reflect the merging of mechanical and electrical engineering disciplines.

coin 作动词用，意为"创造"；to reflect the merging of …是不定式作目的状语。本句可译为：1969 年，安川电气公司的一位日本工程师创造了"mechatronics"（机电一体化）一词，以反映机械与电气工程学科的融合。

2. Evidently, the design and manufacture of future products will involve a combination of precision mechanical and electronic systems, and mechatronics will form the core of all activities in products and production technology.

evidently 是副词在句首作评注状语，意为"显而易见"的；will form the core of 此处等于 turn to be 或 become。

本句可译为：显然，未来产品的设计和制造将会涉及精密机械与电子系统的结合，机电一体化将成为所有产品和生产技术相关活动的核心。

Learn and Practice

1. Mark the following statements with T (true) or F (false) according to the text.

1) An engineer coined the term "mechatronics" in 1968 to reflect the merging of mechanical and electrical engineering disciplines. (   )

2) Now mechatronics still mean a mechanism that is electrified. (   )

Unit 7
Mechatronics Technology

3) Mechatronics emphasizes on the design of robots. (   )

5) It is obvious that mechatronics will gradually become the core of activities in the future manufacturing technology. (   )

2. Choose the best choices according to the text.

1) Mechatronics is the interdisciplinary fusion of mechanics, (   ) and information technology.
   A. physics   B. biology   C. electronics

2) A Japanese engineer (   ) the term "mechatronics" in 1969 to reflect the merging of mechanical and electrical engineering disciplines.
   A. used   B. coined   C. discovered

3) The term encompasses a large array of technologies, many of (   ) have become well-know in their own right.
   A. what   B. which   C. where

4) Many early robots resulting (   ) mechanical and electrical systems became central to mechatronics.
   A. in   B. of   C. from

3. Translate the following phrases into English or Chinese.

1) 机械工程。
2) 机电一体化。
3) 微电子技术。
4) information technology.
5) power electronic technology.

Extensive Reading

PLC Overview

Programmable Logic Controller (*PLC*) (Fig. 7-3), also referred to as programmable controller, is in the computer family. They are used in commercial and industrial applications. A PLC *monitors* inputs, makes decisions based on its program, and controls outputs to automate a process or machine.[1] This article is meant to supply you with basic information about the functions and *configurations* of PLC.

PLC generally consists of input modules or points, a Central Processing Unit (CPU), and output modules or points (Fig. 7-4). An input accepts a variety of *digital or analog signals* from various field devices (sensors) and converts them into logic signals that can be used by the CPU.[2] The CPU makes decisions and executes control instructions based on program instructions in memory. Output modules convert control instructions from the CPU into a digital or analog signal that can be used to control various field devices (*actuators*). A programming device is used to input the desired instructions. These instructions determine what the PLC will do for a specific input. An operator

Fig. 7-3 Programmable Logic Controller

interface allows process information to be displayed and new control parameters to be entered. Pushbuttons (sensors), in this simple example, connected to PLC inputs, can be used to start and stop a motor connected to a PLC through a motor *starter* (actuator).

Prior to PLC, many of these control tasks were solved by contactors or *relay* control. This is often referred to as *hardwired control*. Circuit diagrams had to be designed, electrical components specified and installed, and wiring lists created. Electricians would then wire the components necessary to perform a specific task. [3] If an error was made the wires had to be *reconnected* correctly. A change in function or system expansion required extensive component changes and rewiring.

Fig. 7-4 PLC control system

The same, as well as more *complex* tasks can be done with a PLC. Wiring between devices and relay contacts is done in the PLC program. Hard wiring, though still required to connect field devices, is less intensive. Modifying the application and correcting errors are easier to handle. It is easier to create and change a program in a PLC than it is to wire and rewire a circuit.

Following are just a few of the advantages of PLC:

1) Smaller physical size than hard wire solutions.

2) Easier and faster to make changes.

3) PLC has integrated *diagnostics* and *override* functions.
4) Applications can be immediately documented.
5) Applications can be duplicated faster and less expensively.

Words and Expressions

programmable ['prəugræməbl] adj. 可设计的，可编程的
monitor ['mɔnitə] n. 显示屏，屏幕；监测仪
configuration [kənˌfigju'reiʃən] n. 组合；构造；组态
actuator ['æktjueitə] n. 激励者；执行机构；螺线管
starter ['stɑːtə] n. （发动机的）起动装置；参赛人
relay ['riːlei] n. 接替人员；传递；继电器
reconnect [ˌriːkə'nekt] vt. 再供应，再接通
complex ['kɔmpleks] adj. 复杂的；合成的
diagnostics [ˌdaiəg'nɔstiks] n. 诊断学
override [ˌəuvə'raid] vt. 推翻；优先于；覆盖
Programmable Logic Controller (PLC) 可编程逻辑控制器
digital signals 数字信号
analog signals 模拟信号
hardwired control 硬接线控制

Special Difficulties

1. A PLC monitors inputs, makes decisions based on its program, and controls outputs to automate a process or machine.

to automate a process or machine 是不定式作目的状语；based on its program 是过去分词短语作后置定语，修饰 decisions，表示决策是"根据程序"得到的。

本句可译为：PLC 通过监控输入、根据程序做出决策、控制输出来自动控制过程或机器。

2. An input accepts a variety of digital or analog signals from various field devices (sensors) and converts them into logic signals that can be used by the CPU.

and 连接 accepts 和 converts 引导的两个并列句。

本句可译为：输入单元可以接收来自不同现场设备（传感器）的各种数字或模拟信号，然后将它们转换为 CPU 可以使用的逻辑信号。

3. Circuit diagrams had to be designed, electrical components specified and installed, and wiring lists created. Electricians would then wire the components necessary to perform a specific task.

本句可译为：硬接线控制必须先设计电路图，确定和安装电气元件，创建接线表，然后电工再连接执行特定任务需要的元件。

Learn and Practice

1. Mark the following statements with T (true) or F (false) according to the text.

1) PLC used in commercial and industrial applications. ( )

2) Circuit diagrams had to be designed, electrical components specified and installed. ( )

3) The PLC program is very easy to modify. ( )

2. Work with your partner to answer the following questions

1) What is PLC?

2) What are the advantages of PLC? Give the details.

3) How does a PLC work?

3. Translate the following sentences into Chinese.

1) A super capacitor, so named because of its ability to maintain a charge for a long period of time, protects data stored in RAM in the event of a power loss.

2) Output devices, such as relays, are connected to the terminal strip under the top cover of the PLC.

3) High-speed instruction allows for events and interrupts to occur independent of the PLC scan time.

Knowledge Link

科技英语的表达方式

一、科技英语的语法特点

科技英语文章一般重在客观地叙述事实，力求严谨和清楚，避免主观成分和感情色彩，这就决定了科技英语文体具有以下语法特点：

（1）时态　时态形式比较单一。

（2）语态　经常使用被动语态，而且没有行为的执行者。

（3）定语　经常使用名词作定语以取得简洁的表达效果。例如，radar range finder target selector switch 表示"雷达测距目标选择开关"。

（4）动词非限定形式　经常使用其扩展句子，包括以下具体形式：

1）动词不定式短语。

2）动名词短语。

3）分词短语和独立分词结构。

（5）名词化　以名词为中心词构成短语以取代句子。

例如，when the experiment has been completed 可改写成名词短语 on completion of the experiment。

(6)多重复合句　长复合句较多，句子中往往嵌套子句。

例如：The simple fact shows that the more of the force of friction is got rid of, the farther will the ball travel, and we are led to infer that, if all the impeding forces of gravitation and resistance could be removed, there is no reason why the ball, once in motion, should ever stop.

本句可译为：这一简单事实表明，摩擦力越小，球滚得越远。由此可以推断，如能除去一切起阻碍作用的引力和阻力，球一旦运动就不会再停下来。

(7)逻辑词语　使用很频繁，可明确表示文章内容的内在联系，有助于清楚地叙述、归纳、推理、论证和概括。

例如 hence, consequently, as a result, however, nevertheless, on the contrary, in short, as mentioned above 等。

(8)叙述方式　常避免用第一人称单数，而用第一人称复数 we 或用 the author（paper, article）等第三人称形式。

例如：The author is particularly grateful to Prof Wang for helpful discussion. We have decided to repeat the experiment.

二、科技英语的时态和语态

在科技英语文章中最常用的动词时态有 5 个，即一般现在时、现在进行时、一般过去时、一般将来时和现在完成时。其中，一般现在时、一般将来时和现在完成时要比另外两种更常用些。但它们的用法不如普通英语那样丰富多样。

1. 一般现在时

一般现在时是科技英语文章中最常见的时态，主要有三种用法：

(1)叙述过程

例如：A scientist observes carefully, applies logical thought to his observations and tries to find relationships in data, etc.

(2)叙述客观事实或科学定理

例如：

1) Sound travels through the air in waves.

2) A complete rotation of the earth takes four minutes short of twenty-four hours.

3) Work is equal to the product of force and the distance through which the force moves.

(3)通常或习惯发生的行为

例如：Alternating current is usually supplied to people's houses at 50 cycles per second.

2. 一般将来时

一般将来时表示将来会发生的行为或情况。口语中常用的 be going to do 的形式在文章中很少用，而用 will, is to, is about to 等。例如，Many things now unknown will become known.

3. 现在完成时

现在完成时表示到现在为止发生的行为，或者已经发生但对现在有影响的行为。例如：

1) During the past few years, several countries have pooled their resources in order to carry out certain types of scientific investigation more efficiently.

2）Cast iron is more brittle than iron which has been heated but not to be melted, but it is easier to shape, as it can be poured into moulds.

4. 被动语态

被动语态在科技英语文章中用得十分频繁，主要有两个原因。第一个原因是科技文章描写行为或状态本身，由谁或由什么作为行为或状态的主体不重要，或者状态的主体没有必要指出，甚至根本指不出来。被动语态使用频繁的第二个原因是便于向后扩展句子，而不至于使句子显得头重脚轻。

例如：In the digital computer the numbers to be manipulated are represented by sequences of digits which are first recorded in suitable code, then converted into positive and negative electrical pulses, and stored in electrical or magnetic registers.

在上面这个句子里，sequences of digits 是被动行为的主体，which 连接的定语从句作为它的定语。如果把这个被动句变成主动句，将会影响句子的表达效果和准确含义。

三、科技英语的常用句型

1. 表示原因

1）Because/Since silver is a very good conductor of electricity, it is widely used in industry.

2）In view of the fact that/On account of the fact that/Owing to the fact that/Seeing that silver is a very good conductor of electricity, it is widely used in industry.

3）The reason why it is widely used in industry is that silver is a very good conductor of electricity.

2. 表示结果

1）The temperature of the gas rises. Consequently/Therefore/As a result/Hence it expands in the cylinder.

2）After burners have to be used. Consequently/Therefore/As a result/Hence fuel consumption is heavier.

3）The aircraft speed is limited. Consequently/Therefore/As a result/Hence it will soon become obsolete.

3. 表示比例关系

1）The professor and student ratio is 1∶10.

2）The air and fuel ratio is 15∶1.

3）The ratio of the clearance volume to the swept volume in a cylinder differs in different type of engine.

4. 表示组成和包含关系

1）The alloy contains 5% nickel and 5% iron.

2）The class consists of twenty-four students.

3）The factory produces components for aircraft.

5. 表示能够、许可

1）The microscope enables scientists to examine very small objects.

2）Expansion joints permit/allow the pipes to expand or contract.

Unit 7
Mechatronics Technology

3) Weakness in the metal <u>caused</u> it <u>to</u> fracture under tension.

6. 表示对比关系

1) At high speed the turbo-jet is more efficient, <u>while/whereas</u> at low speed the propeller is more efficient.

2) A hot engine will run on a weak mixture, <u>while on the other hand</u> a cold engine requires a richer mixture.

7. 表示目的、设想

1) <u>For the purposes</u> of some calculation or experiment, we <u>assume</u> certain facts: they may not be true, but it simplifies the calculation to assume they are true.

2) The cylinder <u>is assumed to be</u> a perfect non-conductor of heat.

3) <u>Assuming that</u> the deflection of the galvanometer is 45 degrees, find the weight of copper deposited.

8. 表示计算、判断

(1) <u>Calculate/Work out/Determine</u> the brake horse power developed by the engine at 2000rpm.

(2) The purpose of the test is to <u>determine</u> the calorific value of the fuel.

(3) The temperature of the metal may <u>be estimated/gauged from</u> the color of the oxide film.

Dialogue Exercise

Background: Li Lei and Han Meimei are having a talk with Mr. Wang on mechatronics.

Li Lei: Mr Wang, I feel puzzled at the term mechatronics. If our specialty becomes what we call mechatronics, may we lose the particular features of our major of mechanics.

Mr. Wang: Oh, that goes too far. Mechatronics is nothing but a manufacturing tendency. It just means a combined method or way to design and produce better products. A good example is the facilities for office automation, from the office computer, printer to the smart looking file cabinet. Everything is made mechatronically.

Li Lei: That's OK. But what is the difference between mechanical products and mechatronical products.

Mr. Wang: It's great difference. Mechatronical products, the computer disk drive, for example, can clearly show us its marvelous features, such as much greater flexibility, easier to redesign and reprogram, etc. Imagine, was it possible for us to make anything of that sort 25 years ago?

Han Meimei: No, impossible. I see what you mean. But if so, do we still need to divide mechanical products and electronic ones?

Mr. Wang: Uh…That's up to the real need. We certainly do not take a car as a product of electronics. Instead, it is mainly a mechanical engineering product, though made as a combined process of mechatronics.

81

科 技 英 语

> **Famous Quotes**
>
> **Successful people are always willing to do unsuccessful people did not want to do.**
> 成功的人总是愿意做没成功的人不愿意做的事。

Unit 8　Multi-sensor Data Fusion

知识目标：

1. 了解多传感器数据融合技术的工作原理及相关专业术语。
2. 掌握多传感器数据融合技术的应用领域。
3. 掌握科技英语词汇的增减翻译及转译方法。

能力目标：

1. 能对多传感器数据融合技术的专业术语进行中英互译。
2. 能对多传感器数据融合技术相关英文资料进行阅读和翻译。
3. 能正确使用科技英语词汇的增减翻译及转译方法。

Reading Material

INTRODUCTION：Multi-sensor data *fusion* (MDF) is an emerging technology to fuse data from multiple sensors in order to make a more accurate estimation of the environment through measurement and detection. Techniques for data fusion are integrated from a wide variety of disciplines, including signal processing, pattern recognition, and statistical estimation, artificial intelligence, and control theory. The rapid evolution of computers, proliferation of microelectrical system sensors, and the maturation of data fusion technology provide a basis for utilization of data fusion in everyday applications.

Multi-sensor data fusion provides an approach improving of performance of single sensor (Fig. 8-1). In general, the data of measurement from a single sensor are limited to achieve a high quality. If a number of sensors can be used to perform the same measurement and the data of measurements from these sensors can be combined some way, the resultant has a great potential to outperform over each single measurement with increased *accuracy*. [1]

In general, as its name implies, MDF is a technique by which data from a number of sensors are combined through a centralized data processor to provide comprehensive and accurate informa-

Fig. 8-1   The diagrammatic sketch of MDF

tion. Applications of MDF cross a wide spectrum, including the areas in *military* services such as automatic target *detection* and tracking, battlefield *surveillance*, etc., and the areas in civilian applications such as environment surveillance and monitoring, monitoring of complex machinery, medical diagnosis, smart building, food quality characterization and even precision agriculture (Fig. 8-2). [2]

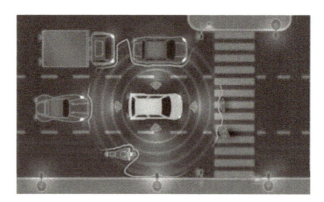

Fig. 8-2   The applications of MDF

In data analysis and processing of measurement and instrumentation, pattern recognition techniques are necessary. Pattern recognition is used to develop data fusion algorithms. Artificial neural networks, which have been developed based on studies about the mechanism of human brain, are the top option over other conventional statistical pattern recognition methods. [3] Linn and Hall *surveyed* more than fifty data fusion systems. Only three of these systems used the neural network method. This low number may indicate an underestimation of the importance of neural networks in the field of data fusion.

Artificial neural networks have been widely used in solving complex problems, such as pattern recognition, fast information processing and adaptation. Artificial neural networks are structured based on studies about the mechanism and the structure of human brain. The architecture and implementation of a neural network models a simplified version of the structure and activities of the human brain. The vast processing power inherent in biological neural structures has inspired the study of the structure itself as a model for designing and organizing man-made computing structure. A MDF *scheme integrated with neural network pattern recognition is a promising structure to achieve high quality data analysis and processing in measurement and instrumentation.*

Words and Expressions

fusion [ˈfjuːʒn] n. 融合；合并
accuracy [ˈækjərəsi] n. 精确（性），准确（性）
military [ˈmɪlətri] adj. 军事的，军用的
detection [dɪˈtekʃn] n. 侦查；检查
surveillance [sɜːˈveɪləns] n. 监视，监督
survey [ˈsɜːveɪ] v. 调查；勘测
scheme [skiːm] n. 计划；体系；阴谋

Special Difficulties

1. If a number of sensors can be used to perform the same measurement and the data of measurements from these sensors can be combined some way, the resultant has a great potential to outperform over each single measurement with increased accuracy.

outperform 意为"做得比……更好，胜过"。

本句可译为：如果利用多个传感器执行相同的测量，再将这些传感器的测量数据以某种方式组合起来，那么结果将很可能超过每个单独测量结果的精度。

2. Applications of MDF cross a wide spectrum, including the areas in military services such as automatic target detection and tracking, battlefield surveillance, etc. and the areas in civilian applications such as environment surveillance and monitoring, monitoring of complex machinery, medical diagnosis, smart building, food quality characterization and even precision agriculture.

such as 意为"例如，譬如"。

本句可译为：MDF 的应用领域广泛，军用方面如目标自动检测与跟踪、战场侦察等；民用方面如环境监测、复杂机械监控、医疗诊断、智能建筑、食品质量检验以及精细农业。

3. Artificial neural networks, which have been developed based on studies about the mechanism of human brain, are the top option over other conventional statistical pattern recognition methods.

本句可译为：人工神经网络，基于对人脑机制的研究而开发，是胜过其他常规统计模式识别方法的首选。

Learn and Practice

1. Mark the following statements with T (true) or F (false) according to the text.

1) MDF is an emerging technology to fuse data from multiple sensors in order to make a more accurate estimation of the environment through measurement and detection. (   )

2) Techniques for data fusion are integrated from a wide variety of disciplines. (   )

3) In general, the data of measurement from a single sensor usually achieve a high quality. (   )

2. Choose the best choices according to the text.

1) Multi-sensor data fusion (   ) an approach improving of performance of single sensor.

A. provides   B. provide   C. provided

2) In data analysis and processing of measurement and instrumentation, pattern recognition techniques are (   ).

A. need   B. necessar   C. necessary

3) Pattern recognition is used to (   ) data fusion algorithms.

A. developing   B. develop   C. developed

4) Linn and Hall (   ) more than fifty data fusion systems.

A. researching   B. surveyed   C. considers

5) The architecture and implementation of a (   ) models a simplified version of the structure and activities of the human brain.

A. neuron   B. neural network   C. neuron cell

3. Translate the following sentences into Chinese or English.

1. MDF 是一项将多个传感器的数据通过一个中央数据处理器进行组合,以提供全面而准确的信息的技术。

2. 人工神经网络已被广泛用于解决复杂问题,例如模式识别、快速信息处理和自适应。

3. Multi-sensor data fusion (MDF) is an emerging technology to fuse data from multiple sensors in order to make a more accurate estimation of the environment through measurement and detection.

4. The vast processing power inherent in biological neural structures has inspired the study of the structure itself as a model for designing and organizing man-made computing structure.

Extensive Reading

Quality Control

According to the American Society for *Quality Control* (ASQC), quality is totality of features and characteristics of a product or service that bear on its ability to satisfy given needs. The definition implies that the needs of the customer must be identified first because satisfaction of those needs is the "bottom line" of achieving quality. Customer needs should then be transformed into product features and characteristics so that a design and the product specifications can be prepared.

Quality is rapidly becoming a major factor in a customer's choice for products and services. The quality control (QC) function has traditionally been performed using *manual inspection methods* and *statistical procedures*. Manual inspection is generally a time-*consuming* procedure which involves *precise*, yet *monotonous* work (Fig. 8-3). It often requires that parts be removed from the *vicinity* of the protection machines to a separate inspection area. This causes delays and often *constitutes* a *bottleneck in the manufacturing schedule.*

Fig. 8-3 Manual inspection for microchip

Inherent in the use of statistical sampling procedures is acknowledgment of the risk that some defective parts will slip through. [1] Indeed, statistical quality control attempts to *guarantee* that a certain expected or average fraction defect rate will be generated during the production inspection process. The nature of traditional statistical QC procedures is that something less than 100% good quality must be tolerated. [2]

Another aspect of the traditional quality control inspection process which *detracts* from its usefulness is often performed after the fact. The measurements are taken and the quality is determined after the parts are already made. If the part is defective, they must be *scrapped* or reworked at a cost which is often greater than their original cost to manufacture. [3]

Quality control includes the activities from the suppliers, through production, and to the customers. *Incoming* materials are examined to make sure they meet the appropriate specifications. The quality of partially completed products is analyzed to determine if production processes are functioning properly. Finished goods and services are studied to determine if they meet customer expectations. With automation, inspection and testing can be so inexpensive and quick that companies may be able to increase sample sizes and the frequency of samples, thus *attaining* more precision in both control charts and acceptance plans.

Words and Expressions

precise [prɪˈsaɪs] adj. 精确的，精密的
bottleneck [ˈbɒtlnek] n. 瓶颈；瓶颈路段
monotonous [məˈnɒtənəs] adj. 枯燥的；(声音) 单调的
vicinity [vəˈsɪnəti] n. 附近；附近地区
constitute [ˈkɒnstɪtjuːt] v. 构成；组成

detract [dɪ'trækt] v. 贬低；减损
guarantee [ˌɡærən'tiː] v. 保证，担保
scrap [skræp] v. 废弃，报废
attain [ə'teɪn] v. 达到，获得
consume [kən'sjuːm] v. 消耗；消费
incoming ['ɪnkʌmɪŋ] adj. 进来的，回来的；新到的
quality control 质量监控
statistical procedures 统计程序
manual inspection methods 手工检测方式

Special Difficulties

1. Inherent in the use of statistical sampling procedures is acknowledgment of the risk that some defective parts will slip through.

inherent in 意为"原有的，固有的"；that 引导同位语从句 some defective parts will slip through，修饰 risk。

本句可译为：使用统计抽样程序的前提是承认存在一些缺陷零件不会被发现的风险。

2. The nature of traditional statistical QC procedures is that something less than 100% good quality must be tolerated.

that 引导表语从句。

本句可译为：传统统计质量控制程序的本质是必须接受某些合格率小于100%的产品。

3. If the part is defective, they must be scrapped or reworked at a cost which is often greater than their original cost to manufacture.

which 引导定语从句对 cost 进行说明。

本句可译为：当某些零件检测不合格时，往往会以高于其原始制造成本的成本报废或返工。

Learn and Practice

1. Mark the following statements with T (true) or F (false) according to the text.

1) According to the text, the quality control (QC) function has only been performed using manual inspection methods. (   )

2) The measurements are taken and the quality is determined at any time. (   )

3) Incoming materials are examined to make sure they meet the appropriate specifications. (   )

4) Finished goods and services are studied to determine if they meet customer expectations. (   )

2. Translate the following passage into Chinese.

With automation, inspection and testing can be so inexpensive and quick that companies may be able to increase sample sizes and the frequency of samples, thus attaining more precision in both control charts and acceptance plans.

Knowledge Link

科技英语词汇的增减翻译及转译

在科技英语翻译过程中，由于英汉两种语言的语法结构的差异、词类差异和修辞手段的不同，往往会出现词义的增补或减少的现象。同时，由于英汉两种语言属于不同语系，同一意思在不同语言中可以用不同词类进行表达。

一、增词译法

增词译法是经常采用的翻译技巧之一，即为了确切表达原文，在译文中增加原文中未出现的某些词。词的增补主要有以下几种情况：重复英语省略的某些词、给动作名词加汉语名词、增加概括性的词、增加解说性的词、增加加强上下文连贯性的词、语法加词、虚拟语气加词、分词独立结构和分词短语的加词等。

1）High voltage is necessary for long transmission while low voltage for safe use.

本句可译为：远距离输电需要高压，安全用电则<u>需要</u>低压。（增加英语句子省略的词）

2）The lower the frequency is, the greater the refraction of a wave will be.

本句可译为：频率越低，波的折射<u>作用</u>就越强。（给动作名词加汉语名词）

3）This report summed up the new achievements made in electron tubes, semiconductors, components.

本句可译为：这篇报告总结了电子管、半导体和元器件<u>三方面</u>的新成就。（增加数量概括词）

4）Many persons learned to program with little understanding of computers or applications to which computer could or should be applied.

本句可译为：许多人<u>虽然</u>已经学会了编程，但对计算机，以及能够或应当使用计算机的应用所知甚少。（增加加强上下文连贯性的词）

5）The development in science to be brought about due to a fuller knowledge of atom is expected to be even more extensive and fundamental.

本句可译为：可以预期，对原子的更加充分的认识所引起的科学发展<u>必将</u>更加广泛，更加基础。（增加词以表示将来）

二、减词译法

减词译法也是英译汉经常采用的翻译技巧之一。在英语中，有些词，如冠词、介词、连词和代词等，在汉语中要么没有，要么用得不多，这主要是因为汉语可以借助词序来表达逻辑关系。因此，词的减少主要出现在英语冠词、介词、代词、连词及关系代词的省略翻译中。

1）<u>The</u> alternating current supplies <u>the</u> greatest part of <u>the</u> electric power for industry today.

本句可译为：如今交流电占了工业用电的绝大部分。（冠词的省略）

2）Because they are neutral electrically, <u>they</u> are called "neutrons".

本句可译为：因为它们呈电中性，所以被称为"中子"。（代词的省略）

3）Because copper possesses good conductivity, it is widely used in electrical engineering.

本句可译为：因为铜具有良好的导电性，所以被广泛地应用于电力工程。（代词的省略）

4）They have found a method for solving this problem.

本句可译为：他们已经找到了解决这个问题的方法。（介词的省略）

5）Check the circuit before you begin in the experiment.

本句可译为：检查好线路再开始做实验。（连词的省略）

三、词性的转译

1. 非动词译成汉语动词

根据汉语动词在使用中的灵活性和广泛性特点，除动词非谓语形式外，还可以把名词、形容词和介词译成汉语动词。

（1）名词译成动词

1）Were there no friction, transmission of motion would be impossible.

本句可译为：没有摩擦就不可能传递运动。

2）The flow of electrons is from the negative zinc plate to the positive copper plate.

本句可译为：电子从负锌极流向正铜极。

（2）形容词译成动词

1）When metal are cut, the shining surface is visible, but it turns gray almost immediately.

本句可译为：切削金属时，可以看到光亮的表面，但几乎立刻就会变成灰色。

2）Copper and gold were available long before man has discovered the way of getting metal from compound.

本句可译为：早在人们找到从化合物中提取金属的方法之前，铜和金就已经开始被使用了。

（3）介词译成动词

1）In general, positive or negative rake tool can be used on stainless steel.

本句可译为：通常，正前角和负前角的刀具都可以用来加工不锈钢。

2）Atomic power for ocean-going vessels is already a reality.

本句可译为：原子能用于远洋船只已经成为现实。

2. 非名词译成汉语名词

（1）动词译成名词

1）A voltmeter connected across A B would read 10Volts.

本句可译为：接在A、B两点间的电压表的读数是10V。

2）Momentum is defined as the product of velocity and a quantity called the mass of the body.

本句可译为：动量的定义是物体的速度和质量的乘积。

（2）形容词译成名词　科技英语往往习惯用表示特征的形容词及其比较级来说明物质的特性，因此在翻译时，可以在这类形容词后加"度""性"等词使之变成名词。

1）The more carbon the steel contains, the harder and stronger it is.

本句可译为：钢的含碳量越高，强度和硬度就越大。

Unit 8
Multi-sensor Data Fusion

2）As most metals are <u>malleable and ductile</u>, they can be beaten into plates and drawn into wire.

本句可译为：由于大多数金属具有<u>韧性和延展性</u>，所以可以将它们压成薄板和拉成细丝。

3. 非形容词译成汉语形容词

（1）副词译成形容词

1）The electronic computer is <u>chiefly</u> characterized by its accurate and rapid computations.

本句可译为：电子计算机的<u>主要</u>特点是计算准确而且迅速。

2）The low cycle fatigue properties are affected <u>adversely</u> by abusive grinding.

本句可译为：过量磨削会对低周疲劳性起到<u>不利的</u>影响。

3）It is a fact that no structural material is <u>perfectly</u> elastic.

本句可译为：事实上，没有一种结构材料是<u>完美的</u>弹性体。

（2）名词译成形容词

1）The low stretches of the rivers show considerable <u>variety</u>.

本句可译为：河流下游的情况是<u>多种多样的</u>。

2）The electrical conductivity has great <u>importance</u> in selecting electrical materials.

本句可译为：导电性在选择电气材料时是很<u>重要的</u>。

Dialogue Exercise

Background: Li Lei and Han Meimei are having some quality control problems. Now, they will review the contract and try to figure it out.

Li Lei: We're having some quality control problems, Han Meimei. We need to go to the source to work them out.

Han Meimei: What are the problems, exactly?

Li Lei: The complaint rate for our new product line is very high, almost seven percent.

Han Meimei: That's high.

Li Lei: Yes. We keep finding problems when testing the boards.

Han Meimei: But your promotional materials claim the boards are the least expensive of their type on the market.

Li Lei: Well, we think that your factory needs to take measures to improve quality control.

Han Meimei: That will involve additional expenses for us, which we'll have to pass on to you as a rate hike.

Li Lei: I'm afraid that's unacceptable. Your contract says that you will deliver a product with a reject rate of less than five percent.

Han Meimei: I'll review the contract and talk with the management. Then we'll get together and figure this out.

> **Famous Quotes**
> **The only limit to our realization of tomorrow will be our doubts of today.**
> 实现明天理想的唯一障碍是今天的疑虑。

Unit 9　Auto-ID Technologies

知识目标：

1. 了解自动识别技术的工作原理和特点。
2. 掌握自动识别技术的应用领域及相关专业术语。
3. 掌握科技英语长句的翻译方法和技巧。

能力目标：

1. 能对自动识别技术的专业术语进行中英互译。
2. 能对自动识别技术相关英文资料进行阅读和翻译。
3. 能正确翻译科技英语中的长句。

Reading Material

INTRODUCTION: RFID is one of *Auto-ID technologies*. Auto-ID technology is anything that collects data about the objects automatically which indentifies and *tracks* faster and *accurately*. In this paper we are going to talk about the Auto-ID technologies, especially the principle and application of RFID.

Auto-ID technology is anything that collects data about the objects into a database without human intervention (Fig. 9-1). [1] Auto-ID technologies are everywhere, quietly and efficiently doing thousands of mundane jobs. The one big job where Auto-ID makes a natural fit is in answering some of the big questions in communication for information like: "What is it?" "Where is it?" and "What about it?"

Primarily the *identification* and tracking of boxes, people, animals, you name it. Compared to humans, Auto-ID technologies identify and track faster, more accurately and reduce overall cost. [2] RFID is only one of many types of Auto-ID technologies. Other Auto-ID technologies include *Magnetic Ink Character Recognition* (*MICR*), magnetic strip, voice recognition, *biometrics* and barcodes.

With all these Auto-ID technologies, why should yet another technology like RFID suddenly becomes so popular? It all boils down to one thing: radio waves. RFID encompasses technologies that

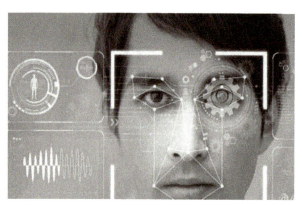

Fig. 9-1　Auto-ID technology

use *electromagnetic* (radio) waves, part of electromagnetic spectrum, to identify *individual* item, place, animal or person (Fig. 9-2).

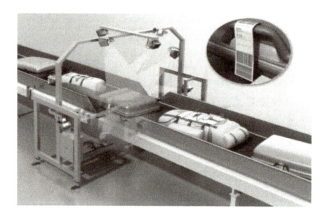

Fig. 9-2　Radio frequency identification (RFID)

RFID can be appropriately implemented for many different uses. The most common is to use an identifying number (sort of name) that uniquely identifies an object, place, animal or person. The number is stored on an integrated circuit that is attached to an antenna. Together, the IC and the antenna are called an RFID transponder or tag. The tag is attached to the object, place, animal or person to be indentified. A device called the interrogator or reader communicates with the tag and it is used to read the identifying number from the tag. The reader feeds the number it reads into an information system, which stores the number in its database or searches its database for the number and returns information stored which is about the object, place, animal or person. The major difference between various Auto-ID technologies is in indentifying number stored and retrieved. [3]

Words and Expressions

primarily [praɪˈmerəli] adv. 根本地；首要地；主要地

identification [aɪˌdentɪfɪˈkeɪʃn] n. 鉴定；识别；验明

Unit 9
Auto-ID Technologies

track [træk] v. 跟踪；检测；追踪
accurately [ˈækjərətli] adv. 正确无误地，准确地
magnetic [mæɡˈnetɪk] adj. 有磁性的；有吸引力的
recognition [ˌrekəɡˈnɪʃn] n. 认识，识别；承认，认可
biometric [ˌbaɪəʊˈmetrɪk] adj. 生物统计学的
electromagnetic [ɪˌlektrəʊmæɡˈnetɪk] adj. 电磁的
individual [ˌɪndɪˈvɪdʒuəl] adj. 个人的；独立的
Auto-ID technology 自动识别技术
MICR (magnetic ink character recognition) 磁墨水字符识别

Special Difficulties

1. Auto-ID technology is anything that collects data about the objects into a database without human intervention.

that 引导定语从句修饰 anything。

本句可译为：自动识别技术是指在没有人为干预的情况下自动收集目标数据、并将数据存入数据库的相关技术。

2. Primarily the identification and tracking of boxes, people, animals, you name it. Compared to humans, Auto-ID technologies identify and track faster, more accurately and reduce overall cost.

本句可译为：自动识别技术主要用于识别和跟踪箱体、人、动物等，但凡能叫出名字的物体都可以识别和跟踪。与人相比，自动识别技术能够更快、更准确，并以更低的总成本进行识别和跟踪。

3. The major difference between various Auto-ID technologies is in indentifying number stored and retrieved.

the major difference 意为"主要区别"。

本句可译为：各种自动识别技术的主要区别为存储和读取识别码的方式不同。

Learn and Practice

1. Mark the following statements with T (true) or F (false) according to the text.

1) Auto-ID technology is anything that collects data about the objects automatically which indentifies and tracks faster and inaccurately. (   )

2) Auto-ID technologies are everywhere, quietly and efficiently doing thousands of mundane jobs. (   )

3) Primarily the identification and tracking of boxes, people, animals, you name it. (   )

4) RFID can be appropriately implemented for many different uses. (   )

5) The IC and the antenna are called an RFID transponder or tag. (   )

2. Choose the best choices according to the text.

1) Auto-ID technologies are everywhere, quietly and (   ) doing thousands of mundane jobs.

A. fast B. efficiently C. slowly

2) RFID encompasses technologies that use electromagnetic (radio) waves, part of electromagnetic spectrum, to identify (   ) item, place, animal or person.

A. individual B. impendent C. single

3) RFID can be appropriately (   ) for many different uses.

A. implemented B. used C. applications

4) A device (   ) the interrogator or reader communicates with the tag and it is used to read the identifying number from the tag.

A. named B. called C. included

3. Translate the following phrases into Chinese or English.

1) Auto-ID technology.
2) RFID.
3) MICR.
4) 数据库。
5) 语音识别。

Extensive Reading

Bioinformatics

Here, we explore a number of issues *elicited* by trying to determine how computing enhances biology and how biology enlivens computer science. [1] How much effort would be expended in redirecting computer scientists to do work in *bioinformatics*? What bioinformatics topics are close to computer science? Whether and how computer science research will find inspiration in biology is long term proposition. [2] Thus, one must probe several related topics, including: the looming *propagation* of biology throughout the sciences; the cultural differences between computer science and molecular biology; the current goals of molecular biology; the web data used in bioinformatics; the areas within computer science of interest to biologists.

Since the 1953 *milestone* achievement by James Watson and Francis Crick determining the structure of DNA, biology, especially molecular biology, has grown by *leaps* and bounds (Fig. 9-3). The sequencing of the human genome represents one of its triumphs. The sequencing of dozens of other *organisms* has followed. Most of these successes would be unthinkable without computers, prompting several questions, including: What is the role of computers in biology? Is it like sending humans to the moon, namely a tool, yet only one among many? Molecular biology is by nature a science of the discrete, a property it shares with computing.

When one field *blends* with another, the usually *hyphenated* term reflecting the combination carries an *ambiguous* meaning. For example, does bio-physics belong to the corpus of knowledge of physics or biology? Is it an independent new discipline? Probably with time and success, the new-discipline interpre-

Fig. 9-3　Structure of DNA

tation holds. Mathematical biology requires a great deal more knowledge of mathematics than of biology. Similarly, computational biology is being developed by computer scientists to satisfy the needs of biologists but basically requires extensive knowledge of computer science theory (Fig.9-4).

Fig. 9-4　Bioinformatics survey

Biologists are not particularly interested in computer science theory for solving their day to day problems. The term "bioinformatics" is more *appealing* to biologists, as the "computation" in "computational biology" is not quite on target. [3] Bioinformatics is a developing interdisciplinary science. The *involvement* of other sciences (such as computer science) holds great promise. Regardless of the outcome, computer scientists are sure to benefit from being active and *assertive* partners with biologists.

Words and Expressions

elicit [iˈlɪsɪt] v. 探出；引出

propagation [ˌprɒpəˈgeɪʃn] n. 传播，传输；蔓延

ambiguous [æmˈbɪgjuəs] adj. 含糊的，不明确的

assertive [əˈsɜːtɪv] adj. 观点明确的；坚决主张的

bioinformatics [bi:əʊɪnˈfɔ:mætɪks] n. 生物信息学
blend [blend] v. 混合；掺杂；结合
leap [li:p] n. & v. 跳跃，飞越
hyphenated [ˈhaɪfəneɪtɪd] adj. 带有连字符的
involvement [ɪnˈvɒlvmənt] n. 牵连；参与；加入
appealing [əˈpi:lɪŋ] adj. 吸引人的，令人心动的
organism [ˈɔ:gənɪzəm] n. 有机体；生物体
milestone [ˈmaɪlstəʊn] n. 里程碑；划时代事件

Special Difficulties

1. We explore a number of issues elicited by trying to determine how computing enhances biology and how biology enlivens computer science.

a number of issues 意为"很多问题"。

本句可译为：我们将探究由试图确定计算机技术如何推进生物学发展以及生物学又是如何使计算机科学更有活力而引出的许多问题。

2. Whether and how computer science research will find inspiration in biology is long term proposition.

long-term proposition 意为"长期命题"。

本句可译为：计算机科学研究能否以及如何在生物学中得到灵感仍然是一个长期的命题。

3. The term "bioinformatics" is more appealing to biologists, as the "computation" in "computational biology" is not quite on target.

本句可译为：术语"生物信息学"对生物学家更有吸引力，因为他们对于"计算生物学"中的"计算"一词不那么感兴趣。

Learn and Practice

1. Mark the following statements with T (true) or F (false) according to the text.

1) Biology has developed rapidly since the 1953 milestone achievement by James Watson and Francis Crick determining the structure of DNA. (   )

2) The sequencing of dozens of other organisms has followed. Most of these successes would be thinkable without computers. (   )

3) When one field blends with another, the usually hyphenated term reflecting the combination carries a clear meaning. (   )

4) Biologists are particularly interested in computer science theory for solving their day to day problems. (   )

2. Choose the best choices according to the text.

1) How much (   ) would be expended in redirecting computer scientists to do work in

bioinformatics?

  A. struggle   B. effort   C. fight

  2) Since the (　　) milestone achievement by James Watson and Francis Crick determining the structure of DNA.

  A. 1953   B. 1954   C. 1958

  3) When one field (　　) with another, the usually hyphenated term reflecting the combination carries an ambiguous meaning.

  A. blends   B. mixed   C. enters

  4) Biologists are not particularly (　　) in computer science theory for solving their day to day problems.

  A. liked   B. interested   C. believed

3. Translate the following paragraph into Chinese.

Biologists are not particularly interested in computer science theory for solving their day to day problems. The term "bioinformatics" is more appealing to biologists, as the "computation" in "computational biology" is not quite on target.

Knowledge Link

科技英语的长句翻译 1

一、常用的翻译方法

  由于其内容的特殊性，科技文章中的长句表达在许多方面有别于日常英语，其差别主要表现在句法和词汇上。大量使用长句是英语科技文章的特点之一。英语长句之所以长，主要长在修饰成分上。英语句子的修饰成分主要是名词后面的定语短语或定语从句，以及动词后面或句首的介词短语或状语从句。这些修饰成分可以一个套一个连用，形成长句结构。

  显然，英语的一句话可以表达好几层意思，而汉语习惯用一个小句表达一层意思，一般多层意思要通过几个短句来表达。由此可见，将英语译成汉语时，有必要进行拆句和改变成分顺序，再按汉语习惯重新组句。长句拆译的方法基本上有四种，即顺译法、倒译法、分译法和综合法。下面分别举例说明。

1. 顺译法

  英语长句结构的顺序与汉语相同，即英语长句中所描述的一连串动作是按时间顺序安排的，可以采用顺译法翻译。

  1) Some of these causes are completely reasonable results of social needs. Others are reasonable consequences of particular advances in science being to some extent self-accelerating.

  本句可译为：在这些原因中，有些完全是自然而然的社会需求，而另一些则是由科学在一定程度上自我加速而产生的特定发展的必然结果。

  2) Typically, when a customer places an order, that order begins a mostly paper-based jour-

ney from in-basket to in-basket around the company, often being keyed and re-keyed into different departments' computer systems along the way.

本句可译为：一般情况下，当一位顾客确定了一份订单后，那张订单就会主要以纸质的形式在公司里传阅，在这一过程中会经常被键入和重复键入到不同部门的计算机系统中。

2. 倒译法

所谓倒译法，就是从长句的后面或中间译起，把长句的开头放在译文的结尾。这是由于英语和汉语的表达习惯不同：英语习惯用前置性陈述，先结果后原因；而汉语则相反，一般先原因后结果，层层递进，最后综合。处理这类句子，就要采用倒译法。

1) Additional social stresses may also occur because of the population explosion or problems arising from mass migration movements—themselves made relatively easy nowadays by modern means of transport.

本句可译为：人口猛增或大量人口流动（现代交通工具使这种流动相对容易实现）造成的种种问题也会对社会造成新的压力。

2) There is no agreement whether methodology refers to the concepts peculiar to historical work in general or to the research techniques appropriate to the various branches of historical inquiry.

本句可译为：所谓方法论，是指一般的历史研究中的特有概念，还是指历史探究中各个具体领域适用的研究手段，人们对此意见不一。

3. 分译法

有时英语长句包含多层意思，而汉语习惯用一个短句表达一层意思。为使行文简洁，翻译时可把长句中的从句或介词短语分开叙述，顺序基本不变，保持前后的连贯；但是有时为了语气上的连贯，须加译适当的词语。

1) The loads a structure is subjected to are divided into dead loads, which include the weights of all the parts of the structure, and live loads, which are due to the weights of people and movable equipment, etc.

本句可译为：一个结构承受的载荷可以分为静载荷和动载荷两类。静载荷包括该结构各部分重量引起的载荷，动载荷则是由人员和可移动设备等重量引起的载荷。

2) When error correction is not required, UDP provides unreliable datagram service that enhances network throughout at the host-to-host transport layer.

本句可译为：如果没有要求更改错误，则UDP会提供不可靠的数据服务，从而增加了网络在主机到主机传输层的吞吐量。

4. 综合法

有些长句单独使用以上三种方法都不合适，则可夹顺夹逆，主次分明，综合处理。

1) Noise can be unpleasant to live even several miles from an aerodrome; if you think what it must be like to share the deck of a ship with several squadrons of jet aircraft, you will realize a modern navy is a good place to study noise.

本句可译为：噪声甚至会使住在离飞机场几英里以外的人感到不适。如果你想象一下和几个中队的喷气式飞机同时处于一艘船的甲板上是什么样的情形，那你就会意识到现代海军船舰是研究噪声的理想场所。

2) The super-cooling effects of the cryogenics which convert liquid helium and other gases into

"superfluids" and metals into "superconductors", making them non-resistant to electricity, could change the world in a number of ways.

本句可译为：低温学的过冷效应可将液态氦及某些气体变成"超流体"，将金属变成"超导体"，使它们没有电阻，从而可通过多种方法改变世界。

二、句子成分的转译

英汉两种语言，由于表达方式不尽相同，翻译时往往需要改变原文的语法结构。所用的主要方法除了词类转换之外，还有句子成分的转换。在一定情况下，适当改变原文中的某些句子成分，可达到译文逻辑正确、通顺流畅、重点突出的目的。

1. 介词宾语译成主语

英译汉时，为了符合汉语的表达习惯，有时需要将原文中的介词宾语转换为原来的主语，以使译文重点突出，行文流畅。

1）Rivers differ greatly in character.

本句可译为：各种河流的特点彼此很不相同。

2）Iron comes between manganese and cobalt in atomic weight.

本句可译为：铁的原子量在锰与钴之间。

3）A motor is similar to a generator in construction.

本句可译为：电动机的结构与发电机类似。

2. 动词 have 的宾语译成主语

1）The proton has considerably more mass than the electron.

本句可译为：质子的质量比电子大得多。

2）Levers have little friction to overcome.

本句可译为：杠杆要克服的摩擦力很小。

3）At high temperature the semiconductor has the same conductivity as the conductor.

本句可译为：高温下半导体的导电性与导体相同。

4）Evidently semiconductors have a lesser conducting capacity than metals.

本句可译为：半导体的导电能力显然比金属差。

3. 其他动词的宾语译成主语

除动词 have 的宾语以外，其他及物动词的宾语有时也可以译成主语，译法基本相同，不过动词（省略不译除外）有时需要与主语一起译成定语。

1）We need frequencies even higher than that we call very high frequency.

本句可译为：我们所需要的频率甚至比我们称之为甚高频的频率还要高。

2）Hot-set systems produce higher strengths and age better than cold-set systems.

本句可译为：热固系统比冷固系统的强度高，而且使用寿命更长。

3）This device (FET) exhibits high impedance.

本句可译为：这种器件（场效应晶体管）的阻抗很高。

4）Light beams can carry more information than radio signals.

本句可译为：光束运载的信息比无线电信号运载的信息多。

4. 主语译成定语

翻译时，往往由于主语更换，而将原来的主语译成定语。

（1）形容词译成名词主语　当形容词译成名词并作主语时，原来的主语通常都需要译成定语。

1）In fission processes the fission fragments are very radioactive.

本句可译为：在裂变过程中，裂变碎块的放射性很强。

2）The wings are responsible for keeping the airplane in the air.

本句可译为：机翼的用途是使飞机在空中保持不下坠。

3）The oxygen atom is nearly 16 times as heavy as the hydrogen atom.

本句可译为：氧原子的重量几乎是氢原子的 16 倍。

（2）动词译成名词主语　当动词译成名词并作主语时，原来的主语一般需要译成定语。

1）The earth acts like a big magnet.

本句可译为：地球的作用像一块大磁铁。

2）Mercury weights about 13 times as much as water.

本句可译为：水银的重量约为水的 13 倍。

3）The vertical spindle-drilling machine is characterized by a single vertical spindle rotating at fixed position.

本句可译为：立式钻床的特点是具有一根单独在固定位置上旋转的垂直主轴。

5. 定语译成谓语

（1）动词宾语的定语译成谓语

1）Copper and tin have a low ability of combining with oxygen.

本句可译为：铜和锡的氧化能力低。

2）Neutron has a mass slightly larger than that of proton.

本句可译为：中子的质量略大于质子的质量。

3）A semiconductor has a poor conductivity at room temperature, but it may become a good conductor at high temperature.

本句可译为：在室温下，半导体的导电性差，但在高温下，它将成为良导体。

（2）介词宾语的定语译成谓语　在介词宾语译作主语的同时，有时还需要把该宾语的定语译成谓语，原来的谓语都译成定语。

1）Gear pumps operate on the very simple principle.

本句可译为：齿轮泵的工作原理很简单。

此句中，将介词宾语 principle 译作主语，其定语 very simple 译成谓语，原来的谓语 operate 译成定语。

2）Nylon is produced by much the same process as rayon.

本句可译为：尼龙的生产过程与人造丝大体相同。

3）Radar works in very much the same way as the flashlight.

本句可译为：雷达的工作原理与手电筒极为相似。

4）Though each cam appears to be quite different from the other, all the cams work in a similar way.

Unit 9
Auto-ID Technologies

本句可译为：虽然每种凸轮都<u>大不相同</u>，但所有凸轮的工作原理都相同。

6. 名词的定语译成汉语主谓结构中的谓语

有时出于修辞的目的，将某一名词前面的形容词，即名词的定语与该名词颠倒翻译，译成汉语的主谓结构，在句子中充当一个成分。原来作定语的形容词译成主谓结构中名词的谓语。

1）Among the advantages of numerical control are <u>more</u> flexibility, <u>higher</u> accuracy, <u>quicker</u> changes, and <u>less</u> machine down time.

本句可译为：数控的优点有灵活性<u>更强</u>、精度<u>更高</u>、换刀<u>更快</u>、非加工时间<u>更短</u>。（试比较：在数控的优点中有更强的灵活性、更高的精度、更快的换刀速度、更短的非加工时间。）

2）These pumps are featured by their <u>simple</u> operation, <u>easy</u> maintenance, <u>low</u> oil consumption and durable service.

本句可译为：这些水泵的特点是操作<u>简便</u>、维修<u>容易</u>、耗油量<u>少</u>、经久耐用。

3）Other requirements of the lathe tool are <u>long</u> life, <u>low</u> power consumption, and <u>low</u> cost.

本句可译为：车刀的其他要求是使用寿命<u>长</u>、能耗<u>低</u>并且造价<u>低</u>。

Dialogue Exercise

Background: Li Lei had some questions on packaging inspection. Through the conversation, Li Lei understood the whole packaging inspection steps.

Li Lei: What procedures do you have in packaging inspection?

Han Meimei: The inspection of packaging is the process in which comparison and assessment are made between the packaging features and the standardized requirements. It includes the inspection, measurement and computation of the features of the packaging.

Li Lei: Can you explain it? It sounds a little bit complicated. How exactly should packaging be inspected?

Han Meimei: Yes, sure. Among the features of packaging, security is of essential importance. For dangerous and poisonous goods, the nature and the generally adopted symbol should be marked conspicuously on each package.

Li Lei: I see what you mean.

Han Meimei: Sanitation is another standard to meet in the inspection of packaging. That is to say, packaging must be done according to the sanitation law and meet the sanitation standard.

Li Lei: I guess this is a very important standard to meet with as to packaging.

Han Meimei: Exactly, besides, packaging has to be convenient and safe in delivery, which is the principle of circulation.

Li Lei: What does that mean exactly?

Han Meimei: It means that you had better choose some light material rather than the heavy material as your packaging material so that it is easier to deliver; but at the same time safety must be taken into account, too. It means that you have to choose some strong material and take some measures like putting foam at the bottom of the container to protect the stuff from being damaged.

Li Lei: Amazing. I have never heard of these before.

Han Meimei: And there is the last standard in the inspection of packaging, the principle of economy. That means you have to use less expensive material rather than more expensive material in packaging if possible.

> **Famous Quotes**
>
> **Men love to wonder, and that is the seed of our science.**
> 人们喜欢猎奇，这就是科学的种子。

Unit 10　The Scientific Exploration of Space

知识目标：

1. 了解太空探索技术及相关专业术语。
2. 掌握太空探索工具的特点及应用领域。
3. 掌握科技英语长句翻译方法和技巧。

能力目标：

1. 能对太空探索技术的专业术语进行中英互译。
2. 能对太空探索技术相关英文资料进行阅读和翻译。
3. 能正确翻译科技英语中的长句。

Reading Material

INTRODUCTION: In recent years, a rapidly growing amount of effort has been devoted to the use of high-power rockets to carry instruments up to great heights above the Earth, to launch artificial satellites and deep space probes. There are four main categories of vehicles involved in this work, which has been called space research. From the scientist's point of view, all these vehicles play a valuable part.

During the last few years, a rapidly growing amount of effort has been devoted to the use of high-power rockets to carry *instruments* up to great heights above the Earth, to launch *artificial satellites* and deep space *probes* (Fig. 10-1). We have pointed out how much has been and can still be done from the earth's surface. Why then all these *concentrate* on the use of rockets?[1]

One of the main reasons is that our *atmosphere*, while beneficial for life in general, prevents us from seeing the universe in any but a very restricted range of light, almost entirely confined to visible light and to a relatively restricted range of radio waves, in fact. We must make observations from outside of the atmosphere to study the *ultra-violet light*, *X-rays*, *infrared rays* and all those radio waves that cannot *penetrate* through our atmosphere. With instruments in artificial satellites circulat-

Fig. 10-1 Chang'e-4 probe

ing at heights of over 200 miles such observations can be made. What they will record, we do not know—if we did, it would not be worth going to all this trouble—but there is *scope* here for *astronomical studies* for generations to come. [2]

This is only one of the many major new possibilities for scientific research which are opened up by the development of rocket *vehicles* in the study of the earth's outer atmosphere, in meteorology, in the study of the space between the earth and the planets, and so on. There are four main categories of vehicles involved in this work, which has been called space research. [3] First, there are vertical sounding rockets. Next, we have the artificial satellites revolving round the earth in elliptical paths. If the satellite orbit is very elongated, so that it passes out to distances several times the earth's radius (4000 miles), we have a deep space probe. Probes may be specially directed to pass near the moon, or hit the moon or become satellites of the moon (Fig. 10-2). These are the lunar probes, of which there have been a number of examples recently.

Fig. 10-2 Far side of the moon

From the scientist's point of view, all these vehicles play a valuable part. [4] The value of any particular launching is the success of the experiment concluded, not just the distance reached from

Unit 10

The Scientific Exploration of Space

the earth. Nor is he concerned with putting men in the vehicle, for the instruments can be made to operate automatically and to send back their readings to earth—even over distances of millions of miles—as *coded* radio signals. [5]

Words and Expressions

instrument ['ɪnstrəmənt] n. 仪器；工具；乐器
probe [prəub] n. 探针；探索；探测仪
concentrate [ˌkɒnsn'treɪt] v. 专心；关注
atmosphere ['ætməsfɪə(r)] n. 大气；风格；气氛
penetrate ['penətreɪt] v. 穿透；渗入；洞悉
artificial [ˌɑːtɪ'fɪʃl] adj. 人造的；虚假的；非原产地的
scope [skəup] n. 范围；余地
astronomical [ˌæstrə'nɒmɪkl] adj. 天文学的；极大的
vehicle ['viːəkl] n. 交通工具；传播媒介
code [kəud] v. 将……译成电码，编码；加密
artificial satellites 人造卫星
ultra-violet light 紫外线
X-ray X 射线
infrared rays 红外线

Special Difficulties

1. We have pointed out how much has been and can still be done from the earth's surface. Why then all these concentrate on the use of rockets?

have pointed out 意为"已经指出了"，注意："have/has + 过去分词"为现在完成时，表示到现在已经完成的动作。

本句可译为：我们已经指出了人们在地球表面已经做了多少，还能做多少。那么，为什么所有这些都集中于火箭的运用呢？

2. What they will record, we do not know—if we did, it would not be worth going to all this trouble—but there is scope here for astronomical studies for generations to come.

be worth doing sth 意为"值得做某事"。

本句可译为：我们不知道这些仪器会记录些什么，如果知道的话，就不值得找这么多麻烦了，但是天文学领域有待于以后几代人进行研究。

3. There are four main categories of vehicles involved in this work, which has been called space research.

be involved in sth 意为"包含在其中"；has been called 意为"被称作"，注意："have/has + been + 过去分词"为现在完成时的被动语态。

107

本句可译为：在称为空间探索的工作中，有四种主要运载工具。

4. From the scientist's point of view, all these vehicles play a valuable part.

from sb's point of view 意为"从某人的观点来看"。

本句可译为：从科学家的观点来看，所有这些运载工具都很有价值。

5. Nor is he concerned with putting men in the vehicle, for the instruments can be made to operate automatically and to send back their readings to earth—even over distances of millions of miles—as coded radio signals.

nor is he concerned with sth 意为"他也不关心某事"。注意：当否定词或含有否定意义的词或短语放在句首时，句子采用部分倒装。由 so, neither, nor 开头的句子，表示重复前面句子的部分意思（表示肯定用 so，表示否定用 neither 或 nor）。

本句可译为：科学家也不关心这些运载工具是否载人，因为仪器可以自动操作并将其记录的数据转换成无线电信号，从数百万英里以外送回到地球。

Learn and Practice

1. Mark the following statements with T (true) or F (false) according to the text.

1) We must make observations from inside the atmosphere to study the ultra-violet light, X-rays, infrared rays and all those radio waves. (   )

2) There are four main categories of vehicles involved in this work, which has been called space research. (   )

3) If the satellite orbit is very elongated, so that it passes out to distances several times the earth's radius (4000 miles), we have artificial satellites. (   )

4) The scientist is concerned with putting men in the vehicle. (   )

2. Choose the best choices according to the text.

1) We have (   ) out how much has been and can still be done from the earth's surface.
   A. points   B. point   C. pointed

2) If we did, it would not be worth (   ) to all this trouble.
   A. to go   B. going   C. gone

3) These are the lunar probes, of (   ) there have been a number of examples recently.
   A. that   B. which   C. what

4) Probes may be (   ) directed to pass near the moon, or hit the moon or become satellites of the moon.
   A. specially   B. particularly   C. expressly

3. Translate the following phrases into Chinese or English.

1) 人造卫星。

2) 月球探测器。

3) vertical sounding rockets.

4) deep space probe.

Unit 10
The Scientific Exploration of Space

Extensive Reading

The Shenzhou Spaceship

The Shenzhou *spaceship* has three main *modules*—the orbital module, the reentry module and *propellant* module, plus an extra section (Fig. 10-3). [1]

The *orbital* module is for the *astronaut* to live in and to use in carrying out his scientific experiments during the flight while in the *earth orbit*. The reentry module is where he stays during the launching period and when returning to the Earth, while the propellant module is to offer power for the craft and to control its position, change its orbit and return it to the Earth.

Fig. 10-3 China Shenzhou spaceship

Lying at the *forefront* of the spacecraft, the orbital module is an air proof *columniform* structure equipped with *solar cells* on both sides, and solar sensors with all sorts of antennae on its exterior. The orbiter is fixed with a *docking* port if docking *maneuvers* are required. [2] If no docking is planned an extra segment is attached.

The uniqueness of the craft lies in its orbital module, which has multiple functions: it becomes a sitting room for astronauts during their flight, a testing cabin when they do experiments, a targeting aircraft when they seek to dock to other craft, an air *brake* when astronauts stroll outside, and a hold when the craft is returning.

The Shenzhou spaceship is distinct from the first manned spaceship of the former Soviet Union and the US in terms of its structure. [3] The Vostok spacecraft, the craft that contained Yuri Gagarin, the former Soviet Union astronaut and the first one in human history, and the Freedom 7 Mercury spacecraft, the one that contained Alan Shepard, the first US astronaut, both had two modules, while the Shenzhou spaceship has three. [4]

In the first space flight, Vostok only circled around the earth for one time and the Mercury craft conducted a *suborbital* flight. The Chinese astronaut, in comparison, is due to spend nearly one day along a low earth orbit. The Shenzhou spaceship mainly uses solar cells, which are superior to the power supply systems on the Mercury and Vostok spaceships.

Words and Expressions

spaceship [ˈspeɪʃɪp] n. 宇宙飞船
module [ˈmɒdjuːl] n. 舱；模块
propellant [prəˈpelənt] adj. 推进的　n. 推进物
orbital [ˈɔːbɪtl] adj. 轨道的
astronaut [ˈæstrənɔːt] n. 宇航员，太空人
forefront [ˈfɔːfrʌnt] n. 最前方；活动中心
columniform [kəˈlʌmnɪfɔːm] adj. 圆柱形的
docking [ˈdɒkɪŋ] v. 对接；减少
maneuver [məˈnuːvə] v. 移动；操纵
brake [breɪk] n. 制动器；阻碍
suborbital [sʌˈbɔːbɪtl] adj. 不满轨道一圈的，亚轨道的
earth orbit 环地轨道
solar cell 太阳能电池

Special Difficulties

1. The Shenzhou spaceship has three main modules—the orbital module, the reentry module and propellant module, plus an extra section.

本句可译为：神舟飞船由轨道舱、返回舱和推进舱三个主要的舱体结构组成，另外还有一个附加段。

2. The orbiter is fixed with a docking port if docking maneuvers are required.

fixed with 意为"配备安装"。

本句可译为：轨道舱设有飞船对接操作需要的对接口。

3. The Shenzhou spaceship is distinct from the first manned spaceship of the former Soviet Union and the US in terms of its structure.

be distinct from 意为"与……截然不同"。

本句可译为：神舟飞船在结构上与最早的苏联和美国的载人飞船有很大区别。

4. The Vostok spacecraft, the craft that contained Yuri Gagarin, the former Soviet Union astronaut and the first one in human history, and the Freedom 7 Mercury spacecraft…

the Vostok spacecraft 指"苏联东方号飞船"，the Freedom 7 Mercury spacecraft 指"美国自由水星7号飞船"。

Unit 10
The Scientific Exploration of Space

Learn and Practice

1. Mark the following statements with T (true) or F (false) according to the text.

1) The Shenzhou spaceship has four main modules: the orbital module, the re-entry module, the propellant module, the extra section. (   )

2) The orbital module is not for the astronaut to live in and to use in carrying out his scientific experiments. (   )

3) The orbital module is an air-proof columniform structure lying at the forefront of the spaceship. (   )

4) The Shenzhou spaceship is quit same like former Soviet Union and the US spacecraft. (   )

5) The Shenzhou spaceship can spend nearly one month along a low earth orbit. (   )

2. Try to answer the following questions.

1) How many parts is the Shenzhou spaceship composed of?

2) Can you explain the functions of the three modules respectively?

3) Why does the uniqueness of the craft lie in its orbital module?

4) The Shenzhou spaceship is distinct from first manned spaceship of the former Soviet Union and the US, why?

3. Translate the following passage into Chinese.

The uniqueness of the craft lies in its orbital module, which has multiple functions: it becomes a sitting room for astronauts during their flight, a testing cabin when they do experiments, a targeting aircraft when they seek to dock to other craft, an air brake when astronauts stroll outside, and a hold when the craft is returning.

Knowledge Link

科技英语的长句翻译 2

科技文章为了能严谨地表达复杂的思想，文章中长句较多。如果把一个长句拆分成若干独立的句子，就有可能影响句子之间内在的密切联系。文章的论述性越强，长句用得就越多，句子也越长。有关科技英语的句型特点在前面的单元中已经做了介绍，本单元我们主要讨论科技英语中较为常见的定语从句和状语从句的翻译方法。

一、定语从句的翻译

定语从句是在复合句中起定语作用的主谓结构，它所修饰的是名词、代词或句子。定语从句可由关系代词（who, whom, which, that, so, but）或关系副词（when, where, why, how）引导，这种结构和汉语中修饰语多置于中心词之前的习惯不同。在翻译定语从句的时候，通常把定语翻译后放在先行词前，即采用倒译法；或者采用把从句译为主句的并列结构，即用顺译法翻译；对于特殊的定语从句，也可采用分译法和合译法进行翻译。

1. 限定性定语从句的翻译

（1）倒译法　将定语从句译成汉语后，放在其先行词前，使译文紧凑，符合汉语习惯。这种翻译方法主要适用于短小且与先行词的关系极其密切的定语从句，译后形成汉语中"……的"结构。

1）Objects that do not transfer light cause shadows.

本句可译为：不透光的物体会产生阴影。

2）A computer is a machine whose function is to accept data and process them into information.

本句可译为：计算机是一种能接收数据并把它们处理成信息的机器。

3）"Genetic engineering" is the name used to describe the techniques that scientists have developed to change the genes of living things.

本句可译为："基因工程"这一名称是用来说明科学家所开发的可改变生物基因的技术。

（2）顺译法　科技文献的特点是精确、简练，因此当定语从句过长时，我们通常会保持原句语序，即将定语从句译成汉语后，放置在主句后，与主句形成并列的句式。

1）Another kind of rectifier consists of a large pear-shaped glass bulb from which all the air has been removed.

本句可译为：另一种整流器由一个大的梨形玻璃泡构成，玻璃泡内的空气已被全部抽出。

2）Until then, they will continue to study moon dust cement and other materials that can make it possible for people to live in the severe environment of the moon.

本句可译为：在此之前，他们将继续研究月球尘土水泥及其他建筑材料，使人们可以在月球的恶劣环境下生活。

3）The base is the foundation of all machines and is the part on which all other parts are mounted.

本句可译为：底座是所有机器的基础，其他部件都装在它上面。

以上例句中，关系代词所代表的含义并没有直接译出。其实，还可以将关系代词译为"它""这""这样"来代替先行词。

（3）合译法　将原主句和分句的界限打破，利用从句的关系代词与主句某成分的替代关系，根据整体含义重新组织成汉语单句，从句在其中可充当定语、谓语、主语或状语。

1）For a long time, scientists could find little use for the material which remained after the oil has been refined. Later on they found that it could be turned into plastics.

本句可译为：科学家们曾在很长一段时间内都没有发现炼油残渣有什么用途，后来才发现能将它制成塑料。

2）Will there be a time when a whole organ of the body—a lung, a kidney, the heart can be frozen for later use?

本句可译为：将来是否有一天可以把人体中像肺、肾、心脏这样完整的器官进行冷冻储存以备后用呢？

3）All of the principal units of the lathe are mounted on a bed having ways along which the carriage and tail stock travel.

本句可译为：车床的主要部件都装在具有导轨的床身上，刀架与尾架可沿着该导轨滑动。

2. 非限定性定语从句的翻译

非限定性定语从句与它所修饰的先行词或句子的关系不甚紧密，只起到补充和说明的作用。它经常由关系代词"which"引导，有时也由"when""where"等关系副词引导，并以逗号与其中心词（句）隔开。翻译非限定性定语从句最常采用的方法是顺译法，根据具体情形不同，也可采用倒译法。

（1）顺译法　如前所述，顺译法即在翻译的过程中保持原句的语序，将从句译成与主句具有并列关系的句式。

1) Fig. 9 shows the effect of continuous charging on energy and electrode consumption, which not only eliminates the need for back charging, but also accomplishes refining during the continuous feed period.

本句可译为：图9表示了连续装料对耗电量和电极消耗的影响，连续装料不仅不需要补充装料，而且能在连续装料阶段内完成精炼操作。

2) Power is equal to work divided by the time, as has been said before.

本句可译为：功率等于功除以时间，这在前面已经讲过。

3) When it comes to communication, we already have Traffic Master, which operates from transducers on motorway bridges to gauge the speed of the traffic and warn of blockage ahead.

本句可译为：至于交通通信，交通管理器已经问世，这种装置通过高速路桥上的传感器对过往车辆的行驶速度进行监测，并警示前方拥堵情况。

（2）倒译法　当非限定性定语从句相对短小时，也不妨使用倒译法。

1) The carbon, of which coal is largely composed, has combined with oxygen from the air and formed an invisible gas called carbon dioxide.

本句可译为：煤的主要成分——碳，同空气中的氧气结合，生成一种无形的气体，叫作二氧化碳。

2) For pyrite and pyrrhonists, which have had limited commercial value to date, the desulphurization process offers an economical means of producing both elemental sulphur and high grade iron oxide.

本句可译为：对于迄今只具有有限工业价值的黄铁矿及磁黄铁矿而言，这种脱硫工艺为同时生产元素硫和高级氧化铁提供了一种经济实用的手段。

二、状语从句的翻译

英语中的状语从句也是比较复杂的句子结构，它形式多样，变化繁多。状语从句由从属连词引导，通常用来修饰主句的谓语动词、形容词、副词，在长句中起状语作用。根据其不同的含义，可以分为时间、地点、原因、条件、让步、结果、目的等各种状语从句。在汉语中，状语从句多在主句前面；英语中的状语从句可在主句之前，也可在主句之后，甚至有时放在整个句子当中。汉语中连词用得不多，语句的逻辑关系是靠词序或语序表示的。在翻译时，连词常可省略，这种情况多见于时间状语从句、条件状语从句和原因状语从句。由于英汉两种语言的差异，其复合句的形式有较大不同，因此在翻译时尤其要注意状语从句的位置、连词的译法和状语从句的转译等。

1. 时间状语从句的翻译

（1）译成相应的时间状语并放在句首　不论原文中表示时间的状语从句是前置还是后置，按照汉语的习惯，时间状语从句要放在其主句的前面。

Heat is always given out by one substance and taken in by another when heat-exchange takes place.

本句可译为：热交换发生时，总是由某一物体放出热量，另一物体吸收热量。

（2）译成并列句　连词（as, while, when 等）引导时间状语从句，在主句和从句的谓语动作同时进行时，翻译时可省略连词，译成汉语并列句。

She sang as she prepared the experiment.

本句可译为：她一边唱歌，一边准备实验。

（3）译成条件状语从句　when 等词引导的状语从句，有时从形式上看是时间状语，但从逻辑上判断则具有条件状语连接词的意义。因此，这类时间状语从句往往可转译为条件状语从句。

Our whole physical universe, when reduced to the simplest terms, is made up of two things, energy and matter.

本句可译为：我们的整个物质世界，如果用最简单的话来说，是由能量和物质这两样东西组成的。

2. 地点状语从句的翻译

（1）译成相应的地点状语　一般可采用前置译法，将地点状语从句译在句首。

Make a mark where you have any doubts or questions.

本句可译为：在任何有困惑或疑问的地方做个记号。

（2）译成条件状语从句或结果状语从句　一些以连词 where 或 wherever 引导的状语从句，有时从形式上看是地点状语，但从逻辑上判断则具有条件状语或结果状语的意义，因而可将其译为条件状语从句或结果状语从句。

1）The materials are excellent for use where the value of the work pieces is not high.

本句可译为：如果零件价值不高，最好使用这种材料。

2）Where the volt is too large as unit, we use the millivolt or microvolt.

本句可译为：如果用伏特作为单位太大，我们可用毫伏或微伏。

3. 原因状语从句的翻译

（1）译成表示原因的分句　汉语中常用来表示原因的关联词有"由于""因为"。一般来说，汉语表示原因的分句置于句首，英语则较灵活。但现代汉语受西方语言的影响，表示原因的分句也有放在后面的，这种情况往往含有补充说明的意义。

1）Some sulfuric dioxide is liberated when coal, heavy oil and gas burn, because they all contain sulfuric compounds.

本句可译为：因为煤、重油和天然气都含有硫化物，所以它们在燃烧时会放出一些二氧化硫。

2）As the moon's gravity is only about 1/6 the gravity of the earth, a 200 pounds man weighs only 33 pounds on the moon.

本句可译为：由于月球的引力只有地球引力的六分之一，所以一个体重 200 磅的人在月球上仅有 33 磅重。

（2）译成因果复合句中的主句　实际上这是一种省略连词的译法，把从句译成主句。

<u>Since</u> information is continuously sent into the system as it becomes available, teletext is always kept up-to-date.

本句可译为：可用的资料被不断地输入到系统中，所以图文电视信息总是保持在最新水平。

4. 条件状语从句的翻译

（1）译成表示条件或假设的分句　在汉语中，"只要""要是""如果"等是表示条件的常用关联词。常用来表示假设的关联词则有"如果""要是""假如"等。按照汉语的习惯，不管表示条件还是假设，分句都放在复句的前部；因此英语的条件状语从句在翻译时绝大多数置于句首。

<u>If</u> something has the ability to adjust itself to the environment, we say it has intelligence.

本句可译为：如果某物具有适应环境的能力，我们就说它具有智力。

（2）译成补充说明情况的分句　绝大多数条件状语从句翻译时都置于句首，但少数可译在主句后面，作为补充说明情况的分句。

Iron or steel parts will rust, <u>if</u> they are unprotected.

本句可译为：铁件或钢件是会生锈的，如果不加保护的话。

5. 结果状语从句的翻译

英语和汉语都把表示结果的状语从句置于主句之后；因此翻译这类从句时可采用顺译法，但又不能拘泥于引导结果状语从句的连词（that, so...that 等）的词义，将其一律译为"因而""结果""如此……以致"等，以免译文过于西化。另外，翻译时应少用连词，或省掉连词。

Some people can not accept the idea that animals might have intelligence <u>so that</u> they are even more surprised at the suggestion that machines might.

本句可译为：有些人不能接受动物可能有智力的想法，因此他们对机器可能有智力的设想就更为惊讶。

6. 目的状语从句的翻译

（1）译成表示目的的后置分句　英语的目的状语从句通常位于句末；因此翻译时一般采用顺译法，即译成后置分句。汉语里常用于表示目的的关联词有"以便""以免""使得"等。

A rocket must attain a speed of about five miles per second <u>so that</u> it may put a satellite into orbit.

本句可译为：火箭必须获得大约每秒五英里的速度才能把卫星送入轨道。

（2）译成表示目的的前置分句　汉语里表示目的的分句常用"为了"作为关联词，置于句首，往往有强调的含意。

All the parts for this kind of machine must be made of especially strong materials <u>in order that</u> they will not break while in use.

本句可译为：为了使用时不致断裂，这种机器的所有部件都应该采用特别坚固的材料制造。

7. 让步状语从句的翻译

(1) 译成表示让步的分句 汉语中表示让步的常用关联词有"虽然""尽管""即使"等，让步状语分句一般前置（但现在也逐渐出现后置现象），英语中则比较灵活。

Although planets give off no light of their own, they reflect the light from the sun and look like stars at night.

本句可译为：虽然行星自己不发光，但是它们反射太阳的光线，夜晚时看起来就像星星。

(2) 译成表示无条件的条件分句 汉语里有一种复句，前一分句指明某一方面的一切条件，后一分句说出在任何条件下都有同样的结果。这类复句的前一分句称为无条件的条件分句，通常以"不管""不论""无论""任凭"等作为关联词。英语中有些让步状语从句可以译为汉语中无条件的条件分句。

All science students, no matter whether they should be physicists or chemists, should have a good foundation in basic sciences.

本句可译为：所有理科学生，未来不论他们是物理学家还是化学家，都应该有良好的基础科学的基础。

Dialogue Exercise

Background: Li Lei is an equipment buyer of a machinery company in China. He was sent to talk with Han Meimei, a foreign sales represent, about the details of the inspection and import of the equipments which they had just bought.

Han Meimei: Shall we talk about the matter of inspection?

Li Lei: All right.

Han Meimei: Since this is the first time that you purchase from us, I'd like to listen to your opinion about inspection.

Li Lei: These mechanical products are not allowed to import to our country without our legal inspection.

Han Meimei: Do you mean that the goods to be imported are subject to legal inspection?

Li Lei: Yes, you are right.

Han Meimei: Then how can I arrange for the deal?

Li Lei: You can send in your samples first for our legal inspection. Once approved, the goods can be shipped to our country.

Han Meimei: It will cost a very long time.

Li Lei: The alternative plan is to ship the goods direct to the bonded area while the samples are being inspected.

Unit 10
The Scientific Exploration of Space

Han Meimei: That sounds good, I'll go and arrange for the deal right now.

> **Famous Quotes**
>
> **Great works are performed not by strength, but by perseverance.**
> 完成伟大的事业不在于体力，而在于坚韧不拔的毅力。

Unit 11　Microprocessor

知识目标：

1. 了解微处理器的组成、特点、工作原理。
2. 掌握微处理器技术应用领域及相关专业术语。
3. 掌握科技文献检索方法。

能力目标：

1. 能对微处理器技术的专业技术进行中英互译。
2. 能对微处理器技术相关英文资料进行阅读和翻译。
3. 能准确、快速检索和利用英文科技文献。

Reading Material

INTRODUCTION： *Microprocessor* is an electronic circuit that functions as the central processing unit (CPU) of a computer, providing computational control. [1] It does not contain large amounts of memory or have the ability to communicate with input devices, such as *keyboards*, joysticks, and *mice*, or with output devices, such as monitors and *printers*. Microprocessors are also used in other advanced electronic systems, such as computer printers, automobiles, and jet airlines.

The microprocessor is one type of *ultra-large-scale integrated circuit.* Integrated circuits, also known as microchips or chips, are complex electronic circuits consisting of extremely tiny components formed on a single, thin, flat piece of material known as a semiconductor (Fig. 11-1). Modern microprocessors incorporate as many as ten million transistors (which act as electronic amplifiers, *oscillators*, or, most commonly, switches), in addition to other components such as resistors, diodes, capacitors, and wires, all packed into an area about the size of a postage stamp. [2]

A microprocessor consists of several different sections: the arithmetic/logic unit (ALU) performs calculations on numbers and makes logical decisions; the registers are special memory locations for storing temporary information much as a scratch pad does; *the control unit deciphers* pro-

Unit 11
Microprocessor

Fig. 11-1　Microprocessor

grams; buses carry digital information throughout the chip and computer; and local memory supports on-chip computation (Fig. 11-2). More complex microprocessors often contain other sections, such as sections of specialized memory, called *cache memory*, to speed up access to external data storage devices. Modern microprocessors operate with bus widths of 64 bits (binary digits, or units of information represented as 1s and 0s), meaning that 64 bits of data can be transferred at the same time. [3]

Fig. 11-2　The application of microprocessor

　　Microprocessors are fabricated using techniques similar to those used for other integrated circuits, such as memory chips. Microprocessors generally have a more complex structure than other chips, and their manufacture requires extremely precise techniques. Economical manufacturing of microprocessors requires mass production. Several hundred dies, or circuit patterns, are created on the surface of a silicon wafer simultaneously. Microprocessors are constructed by a process of deposition and removal of conducting, insulating, and semiconducting materials, one thin layer at a time until, after hundreds of separate steps, a complex sandwich is constructed that contains all the interconnected circuitry of the microprocessor.

Words and Expressions

microprocessor [ˌmaɪkrəʊˈprəʊsesər] n. 微处理器
keyboard [ˈkiːbɔːd] n. 键盘
mouse [maʊs] n. 鼠标，鼠标器，mice［复数］
printer [ˈprɪntər] n. 打印机
oscillator [ˈɒsɪleɪtər] n. 振荡器
decipher [dɪˈsaɪfər] v. 译解，解释
ultra-large-scale integrated circuit 超大规模集成电路
control unit 控制器，控制部件
cache memory 高速缓冲存储器，高速缓存

Special Difficulties

1. Microprocessor is an electronic circuit that functions as the central processing unit (CPU) of a computer, providing computational control.

句中 that 引导的定语从句对 electronic circuit 加以详细说明。

本句可译为：微处理器是一个电子电路，其功能相当于计算机的中央处理单元，可提供计算控制。

2. Modern microprocessors incorporate as many as ten million transistors (which act as electronic amplifiers, oscillators, or, most commonly, switches), in addition to other components such as resistors, diodes, capacitors, and wires, all packed into an area about the size of a postage stamp.

in addition to 意为"除……之外"。

本句可译为：除了电阻、二极管、电容和导线等其他元件之外，现代微处理器还集成了多达 1 千万个晶体管（用作电子放大器、振荡器，或最常见的是用作开关），上述所有元件通通被封装在大约一枚邮票大小的区域内。

3. Modern microprocessors operate with bus widths of 64 bits (binary digits, or units of information represented as 1s and 0 s), meaning that 64 bits of data can be transferred at the same time.

本句可译为：现代微处理器是以 64 比特（比特是二进制数，用 1 和 0 表示的信息单元）的总线宽度工作的，这就是说，可同时传输 64 位数据。

Learn and Practice

1. Mark the following statements with T (true) or F (false) according to the text.

1）Microprocessor contains large amounts of memory. （   ）
2）The microprocessor is one type of ultra-large-scale integrated circuit. （   ）
3）A microprocessor consists of several different sections. （   ）

2. Choose the best choices according to the text.

1) Microprocessors are also used in other advanced (　　) systems, such as computer printers, automobiles, and jet airlines.

　　A. digital　　　　　　B. electronic　　　　　　C. operating

2) Microprocessors are (　　) using techniques similar to those used for other integrated circuits, such as memory chips.

　　A. fabricating　　　　B. fabricate　　　　　　C. fabricated

3) Microprocessors generally have a more (　　) structure than other chips, and their manufacture requires extremely precise techniques.

　　A. simple　　　　　　B. complex　　　　　　C. complicate

4) Several hundred dies, or circuit patterns, are created on the surface of a (　　) wafer simultaneously.

　　A. silicon　　　　　　B. carbon　　　　　　　C. crystal

3. Translate the following sentences into Chinese or English.

1. 集成电路，也被称为微型芯片或者芯片，是复杂的电子电路，由单个、薄而平整的半导体材料上形成的极其微小的元件组成。

2. 更复杂的微处理器往往包含其他部件，例如专用存储区，也叫作高速缓冲存储器，可以加速微处理器对外部数据存储设备的访问。

3. It does not contain large amounts of memory or have the ability to communicate with input devices, such as keyboards, joysticks, and mice, or with output devices, such as monitors and printers.

4. Microprocessors are constructed by a process of deposition and removal of conducting, insulating, and semiconducting materials, one thin layer at a time until, after hundreds of separate steps, a complex sandwich is constructed that contains all the interconnected circuitry of the microprocessor.

Extensive Reading

Digital Signal Processing

A digital signal is a language of ones and zeros that can be processed by mathematics. We speak in real world, analog signals. Analog signals are real world signals that we experience everyday, such as sound, light, temperature, and pressure. A digital signal is a numerical representation of the analog signal. It may be easier and more cost effective to process these signals in the digital world. In the real world, we can convert these signals into digital signals through the analog-to-digital converter, process the signals, and if needed, bring the signals back out to the analog world through the digital-to-analog converter.[1]

Signals may be processed using analog techniques (analog signal processing, or ASP), digital techniques (digital signal processing, or DSP), or a combination of analog and digital techniques (mixed signal processing, or MSP).

Digital signal processor is at the center of signal processing (Fig. 11-3). A digital signal processor (DSP) is a type of microprocessor—one that is *incredibly* fast and powerful. With respect to DSP, the factor that *distinguishes* it from traditional computer analysis of data is its speed and efficiency in performing sophisticated digital processing functions such as filtering, FFT analysis, and data *compression* in real time. [2]

Fig. 11-3　Digital signal processor

A DSP is *unique* because it processes data in real time. This real-time capability makes a DSP perfect for applications that cannot tolerate any delays. For example, did you ever talk on a cell phone where two people couldn't talk at once? You had to wait until the other person finished talking. If you both spoke simultaneously, the signal was cut, you didn't hear the other person. With today's digital cell phones, which use DSP, you can talk normally. The DSP processors inside cell phones process sounds so rapidly you hear them as quickly as you can speak, in real time. Here are just some of the advantages of designing with specialized digital signal processor over other microprocessors:

1) Single-cycle multiply-accumulate operations.

2) Real-time performance simulation and emulation. [3]

3) *Flexibility*.

4) Reliability.

5) Increased system performance.

6) Reduced system cost.

The techniques and applications of digital signal processing are expanding at a tremendous rate. With the advent of large scale integrated circuit and the resulting reduction in cost and size of digital components, together with increasing speed, the applications of digital signal processing techniques is growing (Fig. 11-4). Special purpose digital filters can now be implemented at sampling rates in the *megahertz* range. Digital signal processors also form an integral part of many modern radar and *sonar* systems.

Fig. 11-4　DSP application

Words and Expressions

　　incredibly [ɪnˈkredəbli] adv. 难以置信地；极其
　　distinguish [dɪˈstɪŋgwɪʃ] v. 区分，辨别
　　compression [kəmˈpreʃən] n. 压缩；浓缩
　　unique [juˈniːk] adj. 唯一的，仅有的
　　flexibility [ˌfleksəˈbɪlətɪ] n. 柔韧性，灵活性
　　megahertz [ˈmegəhɜːts] n. 兆赫（MHz）
　　sonar [ˈsəʊnɑːr] n. 声呐装置
　　FFT（Fast Fourier Transform Algorithm）快速傅氏变换算法
　　ASP（analog signal processing）模拟信号处理
　　DSP（digital signal processing）数字信号处理
　　MSP（mixed signal processing）混合信号处理

Special Difficulties

　　1. In the real world, we can convert these signals into digital signals through the analog-to-digital converter, process the signals, and if needed, bring the signals back out to the analog world through the digital-to-analog converter.

　　analog-to-digital converter 意为"模数转换器"；digital-to-analog converter 意为"数模转换"。

　　本句可译为：在现实世界中，我们可以通过模数转换器将模拟信号转换为数字信号，然后对数字信号进行处理；如果需要的话，可以用数模转换器将信号转换到模拟世界中去。

　　2. With respect to DSP, the factor that distinguishes it from traditional computer analysis of data is its speed and efficiency in performing sophisticated digital processing functions such as filtering, FFT analysis, and data compression in real time.

with respect to 意为"与……有关",distinguish from 意为"区分、辨别"。

本句可译为:说到 DSP,它与传统计算机数据分析的区别在于,它在实时执行滤波、FFT 分析和数据压缩等复杂信号处理功能时的速度和效率更快、更高。

3. Real-time performance simulation and emulation.

simulation 意为"模拟",一般指通过软件方法模拟微处理器的功能。emulation 意为"仿真",指使用硬件设备仿真器(emulator)监测微处理器的实时运行。

Learn and Practice

1. Mark the following statements with T(true) or F(false) according to the text.

1) Analog signals are real world signals that we experience everyday sound, light, temperature, and pressure. (   )

2) It may be easier to process these signals in the analog world. (   )

3) We can convert digital signals into analog signals through the analog-to-digital converter. (   )

4) Reliability is one of the advantages of designing with specialized. (   )

2. Translate the following passage into Chinese.

The techniques and applications of digital signal processing are expanding at a tremendous rate. With the advent of large scale integration and the resulting reduction in cost and size of digital components, together with increasing speed, the applications of digital signal processing techniques is growing.

Knowledge Link

科技文献检索和有效利用 1

一、什么是科技文献

科技文献作为记录科技信息或知识的物质载体,传递科技信息或知识,是科技进步的阶梯。科技文献记载了专业技术人员的劳动成果,为后人的科学研究提供了基础,人们通过科技文献不断吸取营养,批判地继承前人的经验,开阔眼界,拓宽思路,在已取得成果的技术上提出新问题,得到新结论,从而攀上新的科学技术高峰。不同国家通过科技文献的交流,使科技信息或知识得以广泛传播和充分利用,这充分体现了科技文献的国际性。

科学技术的不断发展,使科技文献的数量不断增加,质量不断提高。同时,科技文献的发展又加速了科学技术的进步,促进了社会的发展。因此,可以将科技文献的数量和质量作为衡量科学技术发展水平和成就的标志。科技文献包括图书、期刊、科技报告、会议文献、专利文献、学位论文、政府出版物、标准文献、产品资料、科技档案等。在信息技术高速发展的今天,如何从海量的科技信息中检索到对自己有帮助的科技文献,是一名工程技术人员必备的基本素质之一。

二、如何有效地进行科技文献检索

一提到信息检索，大多数人马上就会想到"百度""Google"。人们依赖搜索引擎，在网络上自由搜索自己想要的信息。但困惑也随之而来，我们往往被海量的冗余信息所包围，难以辨别和取舍。怎样才能在信息的"汪洋大海"中有效、快速找到自己所需要的专业信息而不受其他无关信息的干扰呢？仅仅依靠免费的网络资源是不够的，这时专业学术资源就能发挥重要作用了。

专业数据库作为面向学术服务的数字化资源，能确保所提供内容的专业性、准确性、可借鉴性和实用性，可以帮助用户进行知识鉴别和甄选，用户可以直接将专业数据库中的知识拿过来借鉴，这也是专业数据库和目前网络中大量免费资源以及大部分搜索引擎所提供信息的最大区别。

专业数据库包括期刊全文数据库、图书数据库、会议论文数据库、学位论文数据库等。我们经常使用的中文数据库有：CNKI 中国期刊全文数据库、维普中文科技期刊数据库和万方中国数字化期刊群；外文数据库有：SCI、EI、ISTP、ISR 等。可以通过大学图书馆链接和注册账号进行专业信息检索和资源下载。

三、常用专业数据库

1. CNKI 中国期刊全文数据库（中国知网）

（1）CNKI 简介　CNKI（China national knowledge infrastructure），中国期刊全文数据库是由清华同方光盘股份有限公司、中国学术期刊（光盘版）电子杂志社、清华同方知识网络集团等单位联合承担建设的中国期刊全文数据库，也是目前世界上最大的连续动态更新的期刊全文数据库。数据库收录国内核心与专业中英文期刊的全文，数据每日更新，收录年限为 1994 年至今。中国期刊全文数据库集题录、文摘、全文文献信息于一体，对用户搜集科研论文资料、查询行业最新信息、借鉴先进研究成果、充实工作报告内容、制定项目规划、开展多媒体教学等都可提供巨大帮助。

（2）检索方法　中国知网网址为 http：//www.cnki.net，网站提供了中国期刊全文数据库、中国博士学位论文全文数据库、中国优秀硕士学位论文全文数据库、中国重要会议论文全文数据库、中国重要报纸论文全文数据库、中国图书全文数据库、中国年鉴全文数据库、中国标准数据库、中国专利数据库等多个数据库，如图 11-5 所示。CNKI 检索界面提供标准检索、高级检索、专业检索、作者发文检索、句子检索等检索方式，主页默认为标准检索。需要精确检索可使用高级检索，进行检索前，首先要选择需要检索的学科领域和数据库，否则系统默认面向所有学科和全部数据库检索。

（3）高级检索　高级检索界面如图 11-6 所示。高级检索具有多项双词逻辑组合检索、双词频控制功能。多项是指可选择多个检索项；双词是指一个检索项中可以输入两个检索词。各个检索项之间可选择"并且""或者""不含"三种逻辑关系，每个检索项中的两个词之间可选择"并含""或含""不含"三种逻辑关系。可以输入检索范围控制条件，如发表时间、文献来源等，同时可以选择结果的匹配方式，包括"模糊""精确"两种。

2. 维普中文科技期刊数据库

（1）维普数据库简介　中文科技期刊数据库是由重庆维普资讯有限公司开发的中文科技

图 11-5 中国知网主页

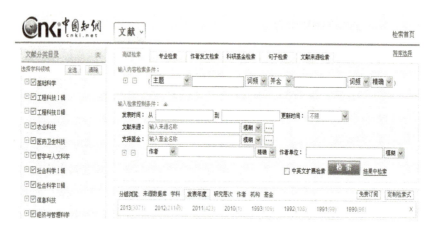

图 11-6 CNKI 高级检索界面

期刊论文数据库，包括全文版、文摘版、引文版三种数据库，是科技查新领域使用最频繁的中文期刊全文数据库。数据库涵盖了数学、经济、化学、生物、农业、环保、地球科学、矿业、机械、无线电、轻工、航空、建筑、医学、教育和图书情报及综合性期刊和港台核心期刊，内容包含标题、作者、机构、刊名（出处）、ISSN 号、CN 号、关键词、分类号、文摘和全文。

（2）检索方法　中文科技期刊数据库检索界面共提供五种检索方式：快速检索、传统检索、高级检索、分类检索和期刊导航，同样也提供全文下载和阅读，用户可以根据需要选择有效的检索方式，如图 11-7 所示。

3. 科技文摘索引数据库

（1）ISI Web of Science　Web of Science 的检索平台为 ISI Web of Knowledge，为用户提供三种检索方式：一般检索（General Search）、引文检索（Cited Reference Search）、高级检索（Advanced Search）。

《科学引文索引》（Science Citation Index，SCI）是由美国科学信息研究所（ISI）创办出

图 11-7　维普数据库检索界面

版的引文数据库，是世界著名的科技文献检索系统之一，是国际公认的进行科学统计与科学评价的主要检索工具。随着技术的发展，SCI 数据库已由原来的光盘数据库发展成为一个网络数据库 SCIE（Science Citation Index Expanded）。SCIE 是一个内容涵盖各学科领域的综合性检索刊物，尤其能反应自然科学研究的学术水平，收录范围是当年国际上的重要期刊，其独特的引文索引法揭示科技文献之间的内在联系，反映文献之间的引用和被引用关系，在学术界占有重要地位。许多国家和地区都将被 SCI 收录及引证的论文情况作为评价学术水平的一个重要指标。SCIE 一般检索界面如图 11-8 所示。

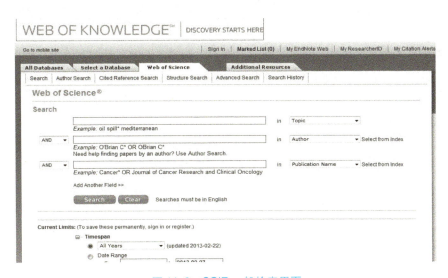

图 11-8　SCIE 一般检索界面

（2）EI Compendex　美国《工程索引》（The Engineering Index，EI）创刊于 1884 年，由美国工程信息公司编辑出版，是查找工程技术领域文献的主要检索工具之一。EI 作为应用科学和工程学在线信息服务的提供者，一直致力于为科学研究者和工程技术人员提供专业化、实用化的在线数据信息服务。EI Compendex 数据库的检索平台为 Engineering Village2

（简称 EV2），该检索平台只使用英语作为检索语言。EV2 平台提供的检索方式包括简单检索、快速检索、专家检索、叙词检索、标签检索。常用的方式是叙词检索、快速检索、专家检索。EI 数据库快速检索界面如图 11-9 所示。

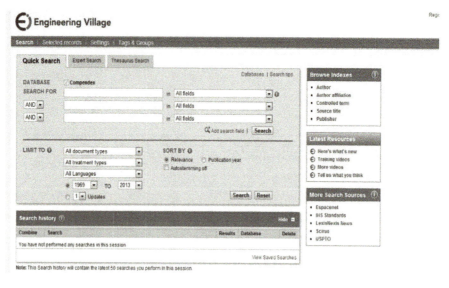

图 11-9　EV2 快速检索界面

Dialogue Exercise

Background：Li Lei and Han Meimei are invited to attend a special talk by Prof. Wang on Computer Integrated Manufacturing. They find the talk very encouraging and helpful for them to learn the future trends of the Manufacturing Industry. Yet, they are not confident whether they can catch the main points of the talk for it covers too many aspects.

Li Lei：Prof. Wang, can I ask a few questions on your talk？

Wang：Sure, go ahead.

Li Lei：Why do you name the manufacturing method computer integrated manufacturing? What do you mean by "integrated"？

Wang：The word "integrated" as an adjective here means "entire or whole". So the phrase CIM means the manufacturing process being controlled by a computer system as a whole. That is, a computer can be used as a controller of everything in manufacturing.

Li Lei：Well, now, I see what you mean. But why do we have to use such a system？

Wang：The aim of its use is first of all to manufacture products more efficiently. With the world economy advancing, the world market will be more and more competitive. In order to win in this market, a manufacturer has to produce the best products at the lowest cost in order to compete with others.

Han Meimei: It's easier said than done. I'm still not clear how manufacturers can win in the world market.

Wang: Oh, don't worry. I think I can answer your question in one sentence. They should try their best to make their products of high volume, large variety but at low costs.

Han Meimei: But how can computer help them to achieve the goal?

Wang: I think computer can help them almost in every way. That is, there will be a thoroughly computer integrated manufacturing system.

Famous Quotes

Science is not for personal glory, not for personal gain, but for human happiness.
科学不是为了个人荣誉，不是为了私利，而是为人类谋幸福。

Unit 12　The Future Car and Automated Driving

知识目标：

1. 了解未来汽车与自动驾驶技术及相关专业术语。
2. 掌握未来汽车技术的发展趋势。
3. 掌握科技文献信息的处理方法。

能力目标：

1. 能对未来汽车与自动驾驶技术的专业术语进行中英互译。
2. 能对未来汽车与自动驾驶技术相关英文资料进行阅读和翻译。
3. 能准确、快速检索和利用英文科技文献。

Reading Material

INTRODUCTION: The modern automobile, evolved from the horse drawn carriage which gradually replaced in the early part of last century. As the technology improves and costs fall, automobile made a place for itself in our daily lives. Now the world auto industry has been developed very fast, each auto manufacturer has tried to win the competition on the world markets and therefore never stopped making research for new models. It seems like we will have *automated driving* and "smart" car in the future.

In the future car we will have switch on automated driving (Fig. 12-1). *ABS* is the first system in the car that can do the *opposite* as the driver commands. [1] ABS reduces the brake power on one wheel against the will of the driver. The result is a safer car by electronic decision-making. The *airbag* decides to inflate by electronic decision-making.

Computer will take over tasks where a simple decision has to be made fast. Today we have the knowledge to get *3D picture* interpreted by a computer: we can measure the distance electronically. The highway can be *overseen* by a computer. The basic tools are available for automated driving on the highway. The problem is that a computer-guided car does not have the right to make mis-

Unit 12

The Future Car and Automated Driving

Fig. 12-1　The automated driving

takes. The manufacture would be *responsible*. So full automated driving will not be for the near future.

　　Safety systems like ABS will come first. Gradually comfort systems like *cruise control* that considers external factors like weather, road marks and traffic around us and *GPS* will enter the car. [2]

　　Automated driving is much more *complex* although we would be happy with a system helping us to avoid driving into the stopping vehicle before us. [3]

Words and Expressions

opposite ['ɒpəzɪt] adj. 相对的，对立的
airbag ['eəbæɡ] n. 安全气囊
oversee [ˌəʊvə'siː] v. 俯瞰；监视
responsible [rɪ'spɒnsəbl] adj. 有责任的，负责的
complex ['kɒmpleks] adj. 复杂的；合成的
automated driving 自动驾驶
ABS 防抱死制动系统
3D picture 三维图片
cruise control 巡航控制
GPS 全球定位系统

Special Difficulties

1. ABS is the first system in the car that can do the opposite as the driver commands.
ABS 是指防抱死制动系统。
本句可译为：防抱死制动系统是汽车上第一个能针对驾驶员命令背道而行的系统。

2. Gradually comfort systems like cruise control that considers external factors like weather, road marks and traffic around us and GPS will enter the car.

gradually 意为"逐步地，渐渐地"。

本句可译为：渐渐地，像巡航控制系统和GPS（全球定位系统）等舒适性系统也将会应用于汽车，巡航控制系统能够适应外部的因素，如天气、路标，以及我们周围的交通状况等。

3. Automated driving is much more complex although we would be happy with a system helping us to avoid driving into the stopping vehicle before us.

although 引导让步状语从句。

本句可译为：尽管我们会乐意有一个系统帮助我们避免类似追尾的交通事故，但自动驾驶技术是非常复杂的。

Learn and Practice

1. Mark the following statements with T (true) or F (false) according to the text.

1) In the future car we will not have switch on automated driving. （　　）
2) The airbag decides to inflate by electronic decision-making. （　　）
3) Today we have the knowledge to get 3D picture interpreted by a computer. （　　）
4) The basic tools are not available for automated driving on the highway. （　　）

2. Choose the best choices according to the text.

1) Computer will take over tasks where a simple (　　) has to be made fast.
   A. decision　　　　B. task　　　　C. code

2) Today we have the knowledge to get 3D picture interpreted by a computer; we can (　　) the distance electronically.
   A. test　　　　B. measure　　　　C. survey

3) The (　　) is that a computer-guided car does not have the right to make mistakes.
   A. question　　　　B. issue　　　　C. problem

4) Automated driving is much more (　　) although we would be happy with a system helping us to avoid driving into the stopping vehicle before us.
   A. difficult　　　　B. complex　　　　C. hard

3. Translate the following phrases into Chinese or English.

1) 自动行驶。　　　　2) 三维图片。
3) cruise control.　　　　4) ABS.
5) GPS.

Extensive Reading

Electric Automobile and Hybrid Power Vehicle

We have also tried steam, gas *turbines*, *rotary* engines and solar powered cars. But none of these could *rival* the power and efficiency of the internal combustion piston engine, fed by *fossil fuel*

Unit 12

The Future Car and Automated Driving

and connected to a transmission that turns the drive wheels through a series of shafts and gears. [1]

Unfortunately, *nagging* problems like fuel shortages and air pollution *necessitate* a renewed search for alternative power sources for our personal transportation needs. While engineers have been *ingenious* in proposing new ideas for *propulsion* systems that are for more efficient than anything that we have today, the problem is that most of these technologies rely on fuels other than gasoline. [2]

It is well know, internal combustion engine automobile pollutant consisting of CO, HC and NO, particle, *ozone* forms the acid rain and *acid fog* and the *photochemical smog* and so on.

Electric vehicles (EVs) are propelled by an electric motor powered by *rechargeable battery packs* (Fig. 12-2). [3] Unlike internal combustion engine automobile, the electric automobile works have not the exhaust pollution. It is extremely beneficial to protect the environmental and the air, so it has *laudatory* title of "Zero-pollution".

Fig. 12-2　The electric automobile

The electric automobile consists of electric drive and the control system, mechanical transmission system, actuator and so on. The electric drive and control system is the electric automobile core, also is most greatly different spot from the internal combustion engine automobile. The electric drive and the control system is composed by the drive motor, the power source and the velocity *modulation* of the electric motor control device and so on. Other basic equipments of the electric automobile are same with the internal combustion engine automobile.

Hybrid electric vehicles (HEVs) provide benefits to consumers, fleets and the nation (Fig. 12-3). These advanced vehicles cut fuel use and costs while maintaining performance, protecting public health and environment, and increasing energy security.

Hybrid electric vehicle emissions vary depending on the vehicle and its configuration. In general, HEVs have lower emissions than conventional vehicles of the same class because the electric motor offsets how much the internal combustion engine is used. [4] In addition, HEVs have the potential to operate in electric—only mode. In this mode, the vehicle operates with no emissions, which is optimal in congested areas and where emissions are not tolerated.

Hybrid electric vehicles need the same general maintenance as conventional vehicles. It is recommended that these vehicles be maintained by a qualified dealer to ensure the *warranty* if maintained.

HEVs are very safe vehicles. They undergo the same *rigorous* testing as conventional vehicles

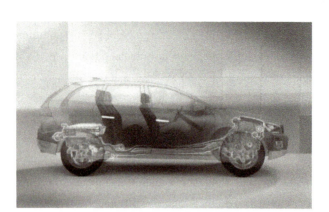

Fig. 12-3　The hybrid electric vehicle

and must meet all of the same standards for safety, including crash testing and airbags.

Words and Expressions

turbine [ˈtɜːbaɪn] n. 涡轮机；汽轮机；叶轮机
rotary [ˈrəʊtəri] adj. 旋转的，转动的；轮流的
rival [ˈraɪvl] v. 竞争；匹敌
nagging [ˈnæɡɪŋ] adj. 挑剔的，纠缠的
necessitate [nəˈsesɪteɪt] v. 需要，迫使
ingenious [ɪnˈdʒiːniəs] adj. 设计独特的，巧妙的；精致的
propulsion [prəˈpʌlʃn] n. 推进，推动力，驱动
ozone [ˈəʊzəʊn] n. 臭氧；新鲜的空气
laudatory [ˈlɔːdətəri] adj. 表扬的，赞扬的
modulation [ˌmɒdjʊˈleɪʃn] n. 调节；调制，调整
hybrid [ˈhaɪbrɪd] adj. 混合的；杂交的
warranty [ˈwɒrənti] n. 理由，根据；授权
rigorous [ˈrɪɡərəs] adj. 严格的，严厉的
fossil fuel 化石燃料，矿物燃料
acid fog 酸雾
photochemical smog 光化烟雾
rechargeable battery pack 可再充电电池组

Special Difficulties

1. But none of these could rival the power and efficiency of the internal combustion piston engine, fed by fossil fuel and connected to a transmission that turns the drive wheels through a series of shafts and gears.

fed by fossil fuel and connected to a transmission 是 the internal combustion piston engine 的后置定

Unit 12
The Future Car and Automated Driving

语，其中 fed 和 connected 是两个并列的过去分词；that 引导的定语从句修饰 transmission。

本句可译为：但所有这些都不能与内燃活塞式发动机的动力和效率相媲美，内燃机由矿物燃料驱动，并与通过一系列轴和齿轮驱动车轮的变速器相连。

2. While engineers have been ingenious in proposing new ideas for propulsion systems that are for more efficient than anything that we have today, the problem is that most of these technologies rely on fuels other than gasoline.

两个 that 从句，第一个修饰 propulsion systems，第二个修饰 anything；other than 意为"不同于；非"。

本句可译为：虽然工程师们一直在潜心研究新的动力系统，使其效率远远超过我们现在使用的动力系统，但问题是，这些技术大都依赖于汽油之外的燃料。

3. Electric vehicles (EVs) are propelled by an electric motor powered by rechargeable battery packs.

powered by…是过去分词结构作定语，修饰 an electric motor。

本句可译为：电动汽车（EVs）由电动机驱动，电动机的动力来自可充电电池组。

4. In general, HEVs have lower emissions than conventional vehicles of the same class because the electric motor offsets how much the internal combustion engine is used.

in general 意为"通常，一般而言"。

本句可译为：一般而言，混合动力电动汽车比同类传统汽车的排放量低，这是因为电动机可一定程度上减少内燃机的使用。

Learn and Practice

1. Mark the following statements with T (true) or F (false) according to the text.

1) The steam, gas turbines, rotary engines and solar powered cars could rival the power and efficiency of the internal combustion piston engine. (   )

2) Internal combustion engine automobile pollutant only consists of CO, HC and NO. (   )

3) Electric vehicles are propelled by an electric motor powered by rechargeable battery packs. (   )

4) Hybrid electric vehicles provide benefits to consumers, fleets and the nation. (   )

5) Hybrid electric vehicles do not need the same general maintenance as conventional vehicles. (   )

2. Choose the best choices according to the text.

1) But none of these could (   ) the power and efficiency of the internal combustion piston engine.

    A. compare   B. rival   C. match

2) Problems like fuel (   ) and air pollution necessitate a renewed search for alternative power sources for our personal transportation needs.

    A. shortages   B. rare   C. lack

3) Electric vehicles are (   ) by an electric motor powered by rechargeable battery packs.

A. promoted   B. propelled   C. drove

4) The electric automobile (　　) of electric drive and the control system, mechanical transmission system, actuator and so on.

A. consists   B. makes up   C. constitutes

5) Hybrid electric vehicles need the same (　　) maintenance as conventional vehicles.

A. common   B. general   C. traditional

3. Translate the following phrases into Chinese or English.

1) 矿物燃料。
2) 可充电电池组。
3) acid fog.
4) photochemical smog.

Knowledge Link

科技文献检索和有效利用2

一、科技文献信息的筛选

随着科学技术的加速发展，信息类型多种多样，信息数量急剧增长，通过不同渠道搜集的信息在质量上存在很大的差异，其中难免存在一定比例的错误、虚假、冗余、过时的信息。这要求我们必须采取必要的措施对信息进行鉴别、筛选，剔除不真实、不准确的信息，使模糊的信息变得清晰，对不完善的信息进行补充，确保去伪存真、去粗取精、系统全面。科技文献检索通常用于科技论文的撰写。在对科技文献信息资料进行筛选时，一般要把握以下几个原则。

1. 可靠性

无论是通过阅读查得的资料，还是在观察、调查、实验中获得的资料，首先都必须保证准确，更不能为了论文的写作而杜撰或伪造资料。不准确的资料应该毫不犹豫地舍弃。一些被怀疑不准确的资料，如果有条件的话，可以重新进行查找或验证；如果不具备重新查找或验证的条件，也应该毫不保留地舍弃。信息内容的可靠性可以经过多方考证，运用累积法和对比分析法可以推断信息的可靠性。

2. 典型性

典型性是指资料能够反映事物的本质，具有强大的说服力。在文献筛选的时候，一定要选择和所研究内容相匹配的资料。任何资料都有典型和非典型之分，一些资料可能在文章甲中属于典型资料，但在文章乙中不属于典型资料。因此，要根据自己的论文情况，选择能够反映论文特色的典型资料。

3. 新颖性

新颖性是指资料具有独创性。也就是自己论文中的资料是别人没有使用过的；或是司空见惯的资料，但做了新的阐释。资料的新颖性往往是文章独创性的一个重要方面，也是论文具有价值的一个重要因素。

4. 充分性

充分性是指资料要足以支撑论文的观点。充分的资料既要能说明论文的观点，使读者对所论述的问题有全面的了解；同时又能表明论文作者的研究水平和创造能力。不同的文章对资料的要求是不同的，而且由于科学的不断发展，新的资料又会不断地出现，因此，要养成积累资料的好习惯。

二、科技文献信息的加工

不是所有搜集到的资料都可以使用，完成资料筛选后，还要对资料进行加工整理。由于资料是支撑论点的重要依据，如果资料出现错误，科技论文的准确性肯定会受到影响。对于那些经过筛选的、参考价值比较高的信息，还要采取一定的方法记录下来，以备日后查找。常见的信息记录方法主要有以下几种。

1. 题录法

题录法就是记录资料的题名、作者、出版者、出版时间等题录信息。例如，对于图书类资料，需记录书名、作者、出版社、出版时间、页码等；对于期刊论文，需记录题名、作者、来源期刊、期号、页码等。

2. 摘录法

摘录法是最简单也最常用的信息记录方法，就是将关键的内容片段摘录下来备用。在阅读资料时，往往会发现一些有用的新观点、新数据等，如果不及时记录下来，以后需要摘引时可能不知从何处查找。因此，在阅读资料过程中应随时将重要内容记录下来，并注明资料来源或出处。

3. 表格法

表格法用于大量零散的、无序的数据记录，并通过统计、计算、推导、折合等方法归纳整理成能说明问题的数据或数据系列。

4. 浓缩法

浓缩法就是对搜集到的原始信息资料进行整理、压缩，择其要点进行记录。信息的浓缩是使信息由博变专、由粗变精的过程，也就是编写文摘的过程。文摘内容要全面，既要忠于原文，摘述文献的观点、目的、研究方法、研究内容、数据、结果和结论等，也要注意简明扼要。编写文摘时还需注意不能遗漏原文的创新性信息。

5. 转化法

转化法就是将不便于直接利用的信息通过改写、编译、翻译等方式转化成便于利用的信息。如在阅读外文资料时，可以将题录和文摘翻译成中文记录，这样以后阅读时就会非常方便。

6. 综述法

综述法就是对某一时期内有关学科、专业、技术或产品所取得的研究成果、所达到的水平以及发展趋势进行综合叙述。

7. 述评法

评述法就是针对某一特定课题，全面搜集国内外有关文献，经过加工、整理、鉴别、分析、综合后，根据国家科技政策和学科理论，进行描述和评论。

三、科技文献信息的分析

1. 比较法

比较法是用对比的方法将两种或两种以上方案、方法、技术、产品等的相应特性或指标，按某种标准或要求，用百分比、基数、图表等进行定量、定性或加以评分比较，以确定其优劣。

常见的比较法有：①求同法，通过调查、考察被研究对象出现或发生的各种场合，寻求研究对象在各种场合出现或发生的共同条件，以确定被研究对象出现或发生的原因。多用于各种情报分析中，分析各种现象出现或发生的因果关系。②求异法，通过调查、考察被研究对象出现或发生的各种场合的不同，确定被研究分析对象出现或发生的原因。③综合比较法，又称全面比较法，从反映事物属性的数量特征、质量特征、结构特征，以及投资成本、规模、品种等各方面进行定性、定量的全面比较。

2. 定性分析法

定性分析法是通过反映事物属性的数量、性能、结构、过程以及诸因素之间的横向关系，进行定性、定量、历史和横断分析，以揭示事物的固有属性和特点。适用于各种综合评论。

常见的定性分析法有：①原理法，是在已经认识了事物的共同本质后，以其为指导，继续对尚未研究的或未深入研究的各种具体事物进行分析，找出其特殊的本质，通过因果关系或内在联系找出结果，得出结论。②典型分析法，是解剖一个典型事例来说明一个观点的分析方法。利用典型分析法可以生动、有力地说明一个问题。这种典型元素可能是有代表性的理论、方案、方法、结构、产品型号、特性等，可用于分析研究和比较认识同类事物的共同本质和属性。③系统分析法，通过对某一课题的调查研究，深入地揭示可能出现的各种情况，并就其经济效果和所涉及的各种问题做详尽的定量、定性分析和比较，在综合分析的基础上提出最佳方案和方法。

3. 定量分析法

定量分析法是指获得研究对象的量的特性的方法。定量分析强调对数据的分析，一般通过建立数学模型进行分析，假设某一事物未来变化量的大小和目前已知的变化规律相同，继续以同样比率沿同一方向或曲线发展变化，因此可以由目前已知的情况推知未来某一时刻或某一条件下的可能情况。

常见的定量分析法有：①文献计量法，采用数学和统计学方法对文献信息进行统计分析，以认识和了解那些表面无序的各类科技数据，揭示和定量预测科学技术的发展和变化。②统计分析法，根据市场情报研究的目的，运用统计综合指标和各种分析法，将已整理的数字资料结合具体情况，由此及彼、由表及里地进行分析、概括，以揭示事物的内在联系及其发展规律。③时间序列法，按照时间顺序排列观察或记录数据序列。是通过研究对象本身时间序列数据的变化，来研究其变化趋势的一种方法。对于搜集到的时间序列数据，一般要分析三种现象：趋向变动、周期变动、不规则变动。

四、科技文献信息的利用

1. 科技文献的典型利用——文献综述

文献综述是作者针对某一专题或领域，对某一时间段内的大量原始研究论文中的数据、

Unit 12
The Future Car and Automated Driving

资料和主要观点进行归纳整理、分析提炼而写成的论文。主要分析和描述前人已经做了哪些工作、进展到何种程度，并对国内外相关研究的动态、前沿性问题做出详细的论述、批判和研究设想，同时提供代表性的参考文献。好的文献综述具有较高的参考价值，通过阅读综述，可以在较短时间内了解当前某一领域中某分支学科或重要专题的最新进展、学术见解和建议。文献综述的特点是要求对文献资料进行综合分析、归纳整理，使材料更精练明确、更有逻辑层次；要求对综合整理后的文献进行专门的、全面的、系统的论述。

2. 合理使用他人的研究成果

科学研究可以利用他人的研究成果，但有个重要的前提：必须遵守知识产权法，遵守学术规范。学术规范是人们在长期的学术实践活动中逐步形成的被学术界公认的一些行为规则。其主要内涵是指在学术活动过程中尊重知识产权和信息道德，严禁抄袭剽窃，在《宪法》《保密法》和《著作权法》的框架下进行学术创作。如果采用了他人文献中的资料和内容，则应注明来源，模型、图表、数据应注明出处。避免"学术不端行为"的有效方法之一是对参考借鉴的文献清楚地指明来源、出处。恰当地引用他人已经公开发表的成果并加以标注，属于合法学术行为，而且科研学者往往欢迎、愿意使自己的科研成果被利用、传播。

Dialogue Exercise

Background: Li Lei has just bought a touch screen mobile phone with large 5.5 HD display. One of his classmates, Han Meimei likes it very much. The following dialogue is happened between them.

Han Meimei: Hi, can I take a look of your new mobile phone? I haven't seen this before.

Li Lei: Sure. It's the Huawei Nova, which is the latest design of their company.

Han Meimei: I like its big touch screen. It's much easier to read messages than mine.

Li Lei: That's one of the reasons why I bought it and the handwriting recognition is fantastic!

Han Meimei: That's really useful, because I type very slowly.

Li Lei: Then it's just right for you. It is an Android smart phone manufactured by Huawei and noted for its combination of a large touch screen, 5.5 inch measured diagonally. It's just like a mini PC.

Han Meimei: Sounds great. I love my LENOVO, but it's a little dated. You know, I may have one of it.

Li Lei: Good idea, but it's too expensive!

Famous Quotes

Do not, for one repulse, give up the purpose that you resolved to effect.

不要只因一次失败，就放弃你原来决心想要达到的目的。

Unit 13　　Computer Network

知识目标：

1. 了解计算机网络的定义及相关专业术语。
2. 掌握计算机网络的应用领域。
3. 掌握科技论文的分类特点和写作要求。

能力目标：

1. 能对计算机网络技术的专业术语进行中英互译。
2. 能对计算机网络技术相关英文资料进行阅读和翻译。
3. 能撰写典型的科技论文。

Reading Material

INTRODUCTION: Computer network is a collection of computers and *devices*, which *facilitates* communications among users and allows resources sharing. Computer network can be used for facilitating communications, sharing hardware, sharing files and sharing *software*. This paper is going to talk about the applications of computer network.

A computer network, often simply referred to as a network, is a collection of computers and devices connected by communications channels that facilitates communications among users and allows users to share resources with other users. [1] A computer network consists of a collection of computers, printers and other equipment that is connected together so that they can communicate with each other. Fig. 13-1 gives an example of a network in a school *comprising* of a local area network (LAN) connecting computers with each other, the Internet, and various servers.

Computer networks can be used for several purposes, as it is shown in Fig. 13-2:

Facilitating communications. Using a network, people can communicate *efficiently* and easily via e-mail, instant messaging, chat rooms, telephone, video telephone calls, and video conferencing.

Sharing hardware. In a networked environment, each computer on a network can access and use

Unit 13
Computer Network

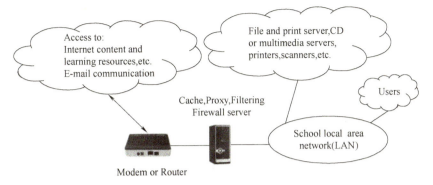

Fig. 13-1 Representation of network in a school

hardware on the network. Suppose several personal computers on a network each require the use of a laser printer. If the personal computers and a laser printer are connected to a network, each user can then access the laser printer on the network, as they need it.

Sharing files, data, and information. In a network environment, any authorized user can access data and information stored on other computers on the network. [2] The *capability* of providing access to data and information on shared storage devices is an important feature of many networks.

Sharing software. Users connected to a network can access *application* programs on the network.

Fig. 13-2 Computer Networks

The connections between computers on a network can be point to point, shared medium, wireless, or a *combination*. The simplest form is point to point, in which a cable connects two computers directly. Messages are passed by writing to the connecting cable and reading from it. More commonly, connections among networked computers require that a communication medium be shared by all the computers on the network. [3] The medium could be a cable or a radio/microware/*satellite transmission* channel.

Words and Expressions

device [dɪˈvaɪs] n. 装置；策略；设备

facilitate [fəˈsɪlɪteɪt] v. 促进；帮助；使容易
software [ˈsɒftweə(r)] n. 软件
comprise [kəmˈpraɪz] v. 包含；由……组成
efficiently [ɪˈfɪʃntlɪ] adv. 有效地；效率高地
capability [ˌkeɪpəˈbɪlətɪ] n. 能力；性能；容量
application [ˌæplɪˈkeɪʃn] n. 应用；申请；用途
combination [ˌkɒmbɪˈneɪʃn] n. 结合；组合；联合
satellite [ˈsætəlaɪt] n. 卫星；人造卫星
transmission [trænzˈmɪʃn] n. 传动装置；传递；传送

Special Difficulties

1. A computer network, often simply referred to as a network, is a collection of computers and devices connected by communications channels that facilitates communications among users and allows users to share resources with other users.

referred to as 意为"被称为……"。

本句可译为：计算机网络，通常简称为网络，是由通信信道连接的计算机和设备的集合，它可促进用户间的通信和资源共享。

2. In a network environment, any authorized user can access data and information stored on other computers on the network.

authorized user 意为"授权用户"。

本句可译为：在网络环境中，任何授权用户都可以访问网络上其他计算机中存储的数据和信息。

3. More commonly, connections among networked computers require that a communication medium be shared by all the computers on the network.

more commonly 意为"更为普遍的"。

本句可译为：更为普遍的是，网络计算机之间的连接要求网络中所有计算机能共享同一通信媒介。

Learn and Practice

1. Mark the following statements with T (true) or F (false) according to the text.

1) In a network environment, any authorized user can not access data and information stored on other computers on the network. (　　)

2) The capability of providing access to data and information on shared storage devices is an important feature of many networks. (　　)

3) Messages are passed by writing to the connecting cable and reading from it. (　　)

4) The connections between computers on a network can be point to point, shared medium, wireless, or a combination. (　　)

2. Choose the best choices according to the text.

1) Computer networks can be used for several purposes, which do not include (   )
   A. facilitating communications   B. sharing files   C. sharing moves

2) A computer network consists (   ) a collection of computers, printers and other equipment.
   A. with   B. of   C. to

3) Users connected to a network can (   ) application programs on the network.
   A. approach   B. enter   C. access

4) Messages are passed by writing to the (   ) cable and reading from it.
   A. connecting   B. connect   C. connected

3. Translate the following phrases into Chinese.

1) share resources.
2) laser printer.
3) sharing hardware.
4) network environment.

Extensive Reading

Information Security

Information security means protecting information and information systems from *unauthorized* access, use, *disclosure*, disruption, modification or destruction.[1] Three basic security concepts important to information on the Internet are *confidentiality*, integrity, and availability (Fig. 13-3). Concepts relating to the people who use that information are *authentication*, authorization, and non-*repudiation*.

When information is read or copied by someone not authorized to do so, the result is known as

Fig. 13-3 Information security

loss of confidentiality.[2] For some types of information, confidentiality is a very important attribute. Examples include research data, medical and insurance records, new product specifications, and corporate investment strategies. In some locations, there may be a legal *obligation* to protect the privacy of individuals. This is particularly true for banks and loan companies; debt collectors; businesses that extend credit to their customers or issue credit cards; hospitals, doctor offices, and medical testing laboratories; individuals or agencies that offer services such as psychological counseling or drug treatment; and agencies that collect taxes.

Information can be *corrupted* when it is available on an insecure network. When information is modified in unexpected ways, the result is known as loss of integrity. This means that unauthorized changes are made to information, whether by human error or intentional tampering. Integrity is particularly important for critical safety and *financial* accounting.

It is remarkably easy to gain unauthorized access to information in an insecure networked environment, and it is hard to catch the intruders (Fig. 13-4).[3] Even if users have nothing stored on their computer that they consider important, that computer can be a "weak link", allowing unauthorized access to the organization's systems and information.

Fig. 13-4 Information leak out

Seemingly *innocuous* information can *expose* a computer system to compromise. Information that intruders find useful includes which hardware and software are being used, system configuration, type of network connections, phone numbers, and access important files and programs, thus compromising the security of system. Examples of important information are passwords, access control files and keys, personnel information, and *encryption* algorithms.[4]

Words and Expressions

unauthorized [ʌnˈɔːθəraɪzd] adj. 未经授权的；未获允许的
disclosure [dɪsˈkləʊzə(r)] n. 公开；泄露
confidentiality [ˌkɒnfɪˌdenʃiˈæləti] n. 机密性

authentication [ɔːˌθentɪ'keɪʃn] n. 鉴定，证明；身份验证
repudiation [rɪˌpjuːdɪ'eɪʃn] n. 拒绝；否认
obligation [ˌɒblɪ'geɪʃn] n. 义务，责任；契约
corrupted [kə'rʌptɪd] v. 败坏；腐化
financial [faɪ'nænʃl] adj. 财政的，财物的
innocuous [ɪ'nɒkjuəs] adj. 无害的；平淡无味的
expose [ɪk'spəuz] v. 揭露，揭发；使暴露
encryption [ɪn'krɪpʃn] n. 编码，加密

Special Difficulties

1. Information security means protecting information and information systems from unauthorized access, use, disclosure, disruption, modification or destruction.

本句可译为：信息安全是指保护信息和信息系统，阻止未经授权的访问、使用、披露、中断、修改或破坏。

2. When information is read or copied by someone not authorized to do so, the result is known as loss of confidentiality.

not authorized 意为 "未经授权的"。

本句可译为：信息被未经授权者读取或复制即构成失密。

3. It is remarkably easy to gain unauthorized access to information in an insecure networked environment, and it is hard to catch the intruders.

remarkably 意为 "明显地，非常地"。

本句可译为：在不安全的网络环境下，未经授权访问信息是非常容易的，而且很难抓住入侵者。

4. Examples of important information are passwords, access control files and keys, personnel information, and encryption algorithms.

本句可译为：如密码、访问控制文件和密钥、人员信息和加密算法等即为重要信息。

Learn and Practice

1. Mark the following statements with T (true) or F (false) according to the text.

1) Two basic security concepts important to information on the Internet are confidentiality and availability. (   )

2) Concepts relating to the people who use that information are authentication, authorization, and non-repudiation. (   )

3) Information can not be corrupted when it is available on an insecure network. (   )

4) Integrity is not important for critical safety and financial accounting. (   )

5) Seemingly innocuous information can expose a computer system to compromise. (   )

2. Choose the best choices according to the text.

1) Three (　　) security concepts important to information on the Internet are confidentiality, integrity, and availability.

  A. basic   B. fundamental   C. elementary

2) Examples include (　　) data, medical and insurance records, new product specifications, and corporate investment strategies.

  A. study   B. research   C. survey

3) Even if users have nothing stored on their computer that they consider important, that computer can be a "(　　)", allowing unauthorized access to the organization's systems and information.

  A. weak link   B. weakness   C. important link

4) Information that intruders find useful may include (　　) important files and programs, thus compromising the security of system.

  A. view   B. scan   C. access

3. Translate the following paragraph into Chinese.

It is remarkably easy to gain unauthorized access to information in an insecure networked environment, and it is hard to catch the intruders. Even if users have nothing stored on their computer that they consider important, that computer can be a "weak link", allowing unauthorized access to the organization's systems and information.

Knowledge Link

科技论文的写作和发表 1

科技论文是对创造性的科研成果进行理论分析和总结的一种文体，是报道自然科学研究和技术开发创新工作成果的论说文章。从内容看，科技论文是创新性科学技术研究工作成果的科学论述，是某些理论性、实验性或观测性新知识的科学记录，是某些已知原理应用于实际中所取得的新进展、新成果的科学总结。从表达方式看，科技论文会运用概念、判断、推理、证明及反驳等逻辑思维手段来分析和表达自然科学理论和技术开发研究成果。

一、科技论文的类型

1. 按写作目的划分

根据写作的目的不同，科技论文可以分为学术性论文、技术性论文、学位论文三种。

学术性论文指研究人员提供给学术性期刊发表或向学术会议提交的论文，以报道学术研究成果为主要内容。

技术性论文指工程技术人员为报道工程技术研究成果而提交的论文，这种研究成果主要是应用已有的理论来解决设计、技术、工艺、设备、材料等领域的具体技术问题而取得的。

学位论文指学位申请者提交的论文，依学位的高低又分为学士论文、硕士论文和博士论文三种。

2. 按研究方式划分

根据研究方式的不同，科技论文可以分为实验研究报告型、理论推导型、理论分析型、设计型、综合论述型五种类型。

实验研究报告型论文追求的是可靠的理论依据、先进的实验设计方案、先进适用的测试手段、合理准确的数据处理及科学严密的分析与论证。其写作重点应放在"研究"上。

理论推导型论文主要是对提出的新的假说进行数学推导和逻辑推理，从而得到新的理论，包括定理、定律和法则。写作时要求数学推导科学、准确，逻辑推理严密，并准确地使用定义和概念，力求得到无懈可击的结论。

理论分析型论文主要是对新的设想、原理、模型、机构、材料、工艺、样品等进行理论分析，对过去的理论分析加以完善、补充或修正。写作时要求论证分析严谨、数学运算正确、资料数据可靠，结论除了准确之外，一般还需经实验验证。

设计型论文的研究对象是新工程、新产品的设计，主要研究方法是对新的最佳设计方案或是实物进行全面论证，从而得出某种结论或引出某些规律。对论文的总体要求是相对要"新"，数学模型的建立和参数的选择要合理，编制的程序要能正常运行，计算结果要合理、准确，设计的产品或调制、配制的物质要经试验证实。

综合论述型论文是作者在博览群书的基础上，综合介绍、分析、评述该学科专业领域国内外的研究新成果、发展新趋势，并表明作者自己的观点，做出发展的科学预测，提出比较中肯的建设性意见和建议。

二、科技论文的特点

1. 创新性

创新性是指科技论文报道的主要研究成果应是前人（或他人）没有提出过的新理论或新知识，要有所发现、有所发明、有所创造、有所前进。

2. 理论性

理论性是指科技论文应具有一定的学术价值，要将实验、观测所得的结果，从理论高度进行分析，把感性认识上升到理性认识，进而找到带有规律性的东西，得出科学的结论。

3. 科学性

科学性包括三个方面的含义：一是指论文内容的科学性，表现为论文内容是真实的，是可以复现的成熟理论、技巧或物件，或者是经过多次使用已成熟、能够推广应用的技术；二是指论文表述的科学性，表现为表述方式准确、明白，语言要十分贴切，如表述概念要进行科学定义或选择恰当的科学术语、表述数字要有准确的数值；三是指论文结构的科学性，表现为论文结构应具有严密的逻辑性，能运用综合方法，从已掌握的材料得出结论。

4. 准确性

准确性是指对研究对象的运动规律和性质表述的准确程度，包括概念、定义、判断、分析和结论要准确，对自己研究成果的估计要确切、恰当，对他人研究成果的评价要实事求是，切忌片面。

5. 规范性

规范性是指科技论文必须按一定格式写作，必须具有良好的可读性。在技术表达方面应符合规范化要求，如名词术语、数字、符号的使用，图表的设计，计量单位的使用，文献的著录等。在文字表达方面，要求语言准确、简明、通顺，条理清楚，层次分明，论述严谨。

三、科技论文的格式和写作要求

1. 科技论文的格式

科技论文一般具有规范的撰写格式要求，其基本结构通常包括前置部分和正文部分。前置部分包括标题、作者、作者单位、摘要、关键词、中图分类号以及英文标题、作者及单位、英文摘要和英文关键词。正文部分包括前言、正文、结论、致谢、参考文献、附录、注释。

2. 标题及写作要求

标题是科技论文的中心和总纲。好的题目要能反映文章的主旨，概括文章的主要内容，起到吸引读者注意力的作用。因此，题名必须确切、鲜明、简洁。

科技文章标题写作应注意以下几点具体要求：

（1）准确得体　题名应能准确地表达论文的中心内容，恰如其分地反映研究的范围和深度，不能使用笼统的、泛指性很强的词语。例如，题名"新能源的利用研究"反映的面很大，而论文实际内容只是研究沼气的利用；题目"煤、电能、劳动力的合理转换"中，概念的外延与内涵不统一，应改为"热能、电能、机械能的合理转换"；有些题名中含有"……的机理""……的规律"一类结构，若其课题的研究深度并不大，则有不注意分寸、有意拔高之嫌，比较客观的做法是取名为"……的一种机制""……现象的解释"。

（2）简短精练　GB/T 7713.1—2006《学位论文编写规则》规定，题名一般不宜超过25个汉字。为了减少题名字数，应尽可能删去多余的词语，避免将同义词或近义词连用。例如，题名"叶轮式增氧机叶轮受力分析探讨"中，"分析"与"探讨"意思接近，可以只保留其一。当题名不易简化时，可加副题名来减少主题名的字数。

（3）便于检索和容易认读　题名所用词语必须有助于选定关键词和编制题录、索引等二次文献，以便为检索提供特定的实用信息。题名中应当避免使用非共知共用的缩略词、首字母缩写字、字符、代号等。

（4）结构应合理　题名尽可能不用动宾结构，而用以名词或名词性词组为中心词的偏正词组。例如，题名"分析研究正压送风量计算方法"可改为"关于正压送风量计算方法的分析研究"，题名"用机械共振法测定引力常数 G"可改为"一种用机械共振法测定引力常数 G 的方法"，题名"研究模糊关系数据库的几个基本理论问题"可改为"模糊关系数据库研究中的几个基本理论问题"。但对于"论……""浅谈……"等形式的题名可用动宾结构，如"试论物流系统的网络模式"。

（5）选词应准确　题名用词应仔细选取，否则会导致语意不明或产生逻辑错误。例如，题名"煎炸油质量测试仪的研制"中，"质量"一词有两种完全不同的含义（一种是物体中所含物质的量，另一种是产品或工作的优劣程度），应改为"煎炸油品质测试仪的研制"。

（6）语序应正确　题名的语序不对，有时造成语意混乱。例如，题名"车辆维修器材计算机信息处理系统"应改为"车辆维修器材管理信息系统的开发"；题名"计算机辅助机床几何精度测试"应改为"机床几何精度的计算机辅助测试"；题名"拱坝的应力特点和分布规律的探讨"应改为"拱坝应力的特点和分布规律的探讨"。

3. 摘要及写作要求

摘要是对论文的内容不加注释和评论的简短陈述，是文章内容的高度浓缩和准确、简洁的摘录。摘要一方面有助于读者尽快了解论文的主要内容以决定是否需要通读该篇论文；另一方面为科技情报人员和计算机检索提供方便。

摘要包括四个方面的内容：一是概述该项研究工作的内容、目的及其重要性；二是介绍所使用的实验方法、实验装置等；三是总结研究成果，突出作者的新见解；四是分析研究结论及其意义。根据摘要内容特点的不同，可以将摘要分为以下三种类型：

一种是报道性摘要，即资料性摘要或情报性摘要。用来报道论文所反映的作者的主要研究成果，向读者提供论文中全部创新内容和尽可能多的定量或定性的信息。尤其适用于试验研究和专题研究类论文，多为学术性期刊所采用。篇幅以 200～300 字为宜。

一种是指示性摘要，即概述性摘要或简介性摘要。它只简要地介绍论文的论题，或者概括地表述研究的目的，仅使读者对论文的主要内容有一个概括的了解。篇幅以 50～100 字为宜。

还有一种是报道—指示性摘要，以报道性摘要的形式表述论文中价值最高的那部分内容，其余部分则以指示性摘要形式表达。篇幅以 100～200 字为宜。

写作摘要时需要注意以下几点：

1）摘要不分段，不列举例证，不描述研究过程。

2）采用第三人称。

3）简短精练，明确具体。摘录出论文的精华，无多余的表述。表意明白，不含糊，无空泛、笼统的词语，应有较多而有用的定性和定量的信息。

4）格式要规范。不用非共知共用的符号和术语，不得简单地重复题名中已有的信息，并切忌罗列段落标题来代替摘要，一般不出现插图、表格以及参考文献序号，不用数学公式和化学结构式。

5）语言通顺，结构严谨，标点符号准确。

4. 前言及写作要求

前言也称"引言""绪论""序"等，是论文中不可缺少的重要部分，目的是向读者交代研究的来龙去脉，其作用在于唤起读者的注意，使读者对论文有一个总体的了解。

前言一般包括三个方面的内容：一是研究的理由、目的和背景，包括问题的提出，研究对象及其基本特征，前人对这一问题做了哪些工作，存在哪些不足，希望解决什么问题，问题的解决有什么作用和意义，研究工作的背景是什么。二是理论依据、实验基础和研究方法。如果沿用已知的理论、原理和方法，只需提及一笔或注出有关的文献；如果要引出新的概念或术语，则应加以定义或阐明。三是预期的结果及其地位、作用和意义。要写得自然、概括、简洁、确切。

写作前言时需要注意以下几点：

（1）言简意赅，突出重点　前言中要求写的内容较多，而篇幅有限，这就需要根据研究课题的具体情况确定阐述重点。

（2）开门见山，不绕圈子　注意一起笔就切题，不能铺垫太多。

（3）尊重科学，不落俗套　例如，有的作者在论文的前言部分总爱对自己的研究工作或能力表示谦虚，寻几句客套话来说，如"限于时间和水平"或"由于经费有限，时间仓促"

"不足或错误之处在所难免,敬请读者批评指正"等。这种情况应当避免。

(4)如实评述 避免吹嘘自己和贬低别人。

5. 正文及写作要求

科技论文的正文部分一般由以下四部分内容组成:

(1)理论分析 理论分析也称基本原理分析,是论证的理论依据,是对论文所做假设及其合理性的阐述,对于分析方法的说明。写作时应注意区别哪些是已知的,哪些是作者首次提出的,哪些是经过改进的,都需要交代清楚。

(2)实验材料和方法的阐述 实验材料的阐述是指对材料的来源、性质和数量,以及材料的选取和处理等事项的阐述。实验方法的阐述是指对实验的仪器、设备以及实验条件和测试方法等事项的阐述。主要内容包括:实验对象,实验材料的名称、来源、性质、数量、选取方法和处理方法,实验目的,使用的仪器、设备(包括型号、名称、测量范围和精度等),实验及测定的方法和过程,出现的问题和采取的措施等。

(3)实验结果及其分析 实验结果及其分析包括给出结果和分析实验所得到的各种现象两部分内容。需要对实验所得结果进行定性或定量的分析并说明其必然性,但不引入前人的研究结果或讨论;其主要内容是以绘图和列表等手段整理实验结果,通过数理统计和误差分析说明结果的可靠性、再现性和普遍性;进行实验结果与理论计算结果的比较,说明结果的适用对象和范围;分析不符合预期的现象和数据,检验理论分析的正确性等。

(4)结果讨论 对结果进行讨论的目的在于阐述结果的意义,说明与前人所得结果不同的原因,根据研究结果继续阐述作者自己的见解。结果的讨论一般包括以下内容:解释所取得的研究成果,说明成果的意义,指出自己的成果与前人研究成果或观点的异同,讨论尚未定论之处和相反的结果,提出研究的方向和问题,最主要的是突出新发现、新发明,说明研究结果的必然性或偶然性。

对正文部分写作的总体要求是论点明确,论据充分,论证合理。文中出现的数据要计算准确,文章在论述过程中条理要清楚、逻辑性要强、表达形式要与内容相适应。在论证过程中要注重科学性,内容要丰富,文字要凝练。对于需要保密的资料一定要做技术处理。

6. 结论及写作要求

结论又称结束语、结语。它是在理论分析和实验验证的基础上,通过严密的逻辑推理得出的富有创造性、指导性、经验性的结果描述。它又以自身的条理性、明确性、客观性反映了论文或研究成果的价值。结论与前言相呼应,同摘要一样,其作用是便于读者阅读和为二次文献作者提供依据。结论不是研究结果的简单重复,而是对研究结果更深入一步的认识,是从正文部分的全部内容出发,并涉及前言的部分内容,经过判断、归纳、推理等过程,将研究结果升华成新的总观点。

结论一般包括三方面的内容:一是本论文的研究结果说明了什么问题,得出了什么规律性的东西,解决了什么实际问题。二是对前人有关本问题的看法做了哪些检验,哪些与本次研究结果一致,哪些不一致,作者做了哪些修正、补充、发展或否定。三是本研究的不足之处或遗留问题。对于某篇论文的结论,上述第一点是必需的,第二点和第三点视论文的具体内容可有可无。如不能导出结论,也可以没有结论而进行必要的讨论。

如果结论段的内容较多,可以分条来写,并进行编号,每条成一段,每段包括一句话或几句话。如果结论段内容较少,则可以写成一段话。结论里应包括必要的数据,但主要进行文字表达，一般不再用插图和表格。

7. 参考文献的著录格式

科研项目的研究都是在前人工作的基础上发展起来的，在学术论文中的具体体现就是参考文献的引用。撰写论文而引用他人文章的论点、材料和结果等，应按照国家标准GB/T 7714—2015《信息与文献　参考文献著录规则》的规定进行标注和著录。

不同的文献类型使用不同的文献类型标识码，例如，普通图书——M、会议录——C、报纸——N、期刊——J、学位论文——D、报告——R、标准——S、专利——P、数据库——DB、计算机程序——CP、电子公告——EB。

常用参考文献的著录格式及示例如下：

（1）普通图书的著录格式　[序号]作者．书名[M]．版次．出版地：出版者，出版年．起止页码．

[1] 王兆安，杨君，刘进军．谐波抑制和无功功率补偿[M]．北京：机械工业出版社，1998．

[2]（美）L·科恩．时频分析：理论与运用[M]．白居宪译．西安：西安交通大学出版社，2000．

（2）期刊论文的著录格式　[序号]作者．文献题名[J]．期刊名称，年，卷（期）：起止页码．

[3] 程明，周鹗，蒋全．双凸极变速永磁电机的静态特性[J]．电工技术学报，1999，14（5）：9-13．

[4] Cheng Ming, Zhou E, Jiang Quan. Static Characteristics of Doubly Salient Permanent Magnet Motors for Adjustable Speed Drives [J]. Transactions of China Electrotechnical Society, 1999, 14 (5): 9-13.

（3）会议录的著录格式　[序号]作者．题名[C]．出版地：出版者，出版年．起止页码．

[5] 钱照明，张军明，谢小高，等．第二届电工技术前沿论坛论文集[C]．北京：电工技术学报编辑部，2005：1-15．

（4）学位论文的著录格式　[序号]作者．题名[D]．保存地点：保存单位，授予年份．

[6] 黄文新．笼型异步发电机——电力电子变换器高压直流发电系统的研究[D]．南京：南京航空航天大学，2002．

（5）报纸的著录格式　[序号]作者．文献题名[N]．报纸名，出版年，版次．

[7] 胡鞍钢．中国能够实现粮食自给目标[N]．联合早报，1994，10．

（6）标准的著录格式　[序号]标准制定者．标准编号和标准名[S]．

[8] 中国汽车技术研究中心．GB/T 4782—2001 道路车辆操纵件、指示器及信号装置词汇[S]．

（7）专利的著录格式　[序号]专利申请者．专利名称：专利国别，专利号[P]．出版日期．

（8）数据库的著录格式　[序号]著者．题名[DB]．[引用日期]．获取和访问路径．

Dialogue Exercise

Background: As a material supplier, Han Meimei visited a local machining enterprise, with the hope of further cooperation. The following conversation happened between Han Meimei and Li Lei who is the owner of this machining enterprise.

Han Meimei: Are there some materials stored for your company's daily use?

Li Lei: There are some material warehouses on the construction site.

Han Meimei: How do you manage them usually?

Li Lei: Our store officer is responsible for the warehousing and issuing of materials. We use scientific-management system for material storage and its control.

Han Meimei: Are these materials bought from the local market?

Li Lei: No, these materials are imported from abroad.

Han Meimei: What are the features about them?

Li Lei: These metal materials fit our purpose satisfactorily.

Han Meimei: Which kind of metal material is on your top consumption?

Li Lei: Typical structural steel shapes include beams, channels, angles and tees.

Han Meimei: Don't you use cast iron in which cost less for manufacturing?

Li Lei: We use them rarely. Cast iron cannot compare with steel in tensile strength.

Han Meimei: Then I think my company can provide materials for your company.

Li Lei: Well, as long as you have the advantage on the transaction price.

Han Meimei: All right, we can compete by cutting prices and making some innovations.

Li Lei: That is perfect. See you later.

Famous Quotes

The important thing in life is to have a great aim, and the determination to attain it.
人生最重要的事情就是确定一个伟大的目标,并决心实现它。

Unit 14　E-commerce

知识目标：

1. 了解电子商务的定义、应用领域及相关专业术语。
2. 掌握电子商务的主要交易类型。
3. 掌握科技论文的选题确定、写作和发表方法。

能力目标：

1. 能对电子商务技术的专业术语进行中英互译。
2. 能对电子商务技术相关英文资料进行阅读和翻译。
3. 能撰写并发表科技论文。

Reading Material

INTRODUCTION：*Electronic Commerce* is commonly known as E-commerce which consists of the buying and selling of products or services over electronic systems such as the Internet and other computer networks. This paper introduces the E-commerce *definition* and types of E-commerce.

E-commerce or electronic commerce is the *purchasing*, selling, and exchanging of goods and services over computer networks (such as the Internet) through which transactions or terms of sale are performed *electronically* (Fig. 14-1). Contrary to popular belief, E-commerce is not just on the Web. In fact, E-commerce was alive and well in business to business transactions before the Web back in the 70s via EDI (Electronic Data Interchange)[1] through VAN (Value Added Network)[2]. E-commerce can be broken into three main categories: B2B, B2C, and C2C.

B2B (Business-to-Business): Companies doing business with each other such as manufacturers selling to distributors and *wholesalers* selling to *retailers*. Pricing is based on quantity of order and is often *negotiable*.

B2C (Business-to-Consumer): Businesses selling to the general public typically through *catalogs* utilizing shopping cart software.

Fig. 14-1　Starting your business with E-commerce

C2C (Consumer-to-consumer): There are many sites offering free classifieds, auctions, and forums where individuals can buy and sell thanks to online payment systems like Alipay where people can send and receive money online with ease. TaoBao is a great example of where person-to-person transactions take place everyday.

G2G (Government-to-Government), G2E (Government-to-Employee), G2B (Government-to-Business), B2G (Business-to-Government), G2C (Government-to-Citizen), C2G (Citizen-to-Government) are other forms of E-commerce that involve transactions with the government—from *procurement* to filing taxes to business registrations to renewing *licenses*. There are other categories of E-commerce out there, but they tend to be *superfluous*.

Fig. 14-2　Mobile payments

Though E-commerce is on Business-to-Business (B2B), Business-to-consumer (B2C), it is being increasingly realized that the B2B holds the most potential. [3] While the global B2C E-commerce grew from $7.80 billions in 1998 to $2.2 *trillions* by 2017, that for B2B grew from $43 billions to $20.3 trillions. Little wonder that the focus world over is now on B2B E-commerce.

Unit 14
E-commerce

Words and Expressions

definition [ˌdefɪ'nɪʃn] n. 定义；规定
purchase ['pɜːtʃəs] v. 购买；换取
electronically [ɪˌlek'trɒnɪkli] adv. 电子地
wholesaler ['həʊlseɪlər] n. 批发商
retailer ['riːteɪlər] n. 零售商，零售店
negotiable [nɪ'gəʊʃiəbl] adj. 可谈判的；可协商的
catalog ['kætəlɔːg] n. 目录；产品样本
procurement [prə'kjʊəmənt] n. 获得，取得；采购
license ['laɪsns] n. 许可证，执照；特许
superfluous [suː'pɜːfluəs] adj. 过多的，多余的；奢侈的
trillion ['trɪljən] n. 万亿，兆
Electronic Commerce 电子商务

Special Difficulties

1. EDI (Electronic Data Interchange)。
在电子商务中指电子数据交换，是电子商务的初级形式。是指企业与企业之间、企业与政府之间通过一个内部网络进行的数据传递和数据交换。

2. VAN (Value Added Network)。
增值网络，是将制造方、批发方、物流方、零售方等各环节的信息，通过计算机服务网络来相互交换的信息系统。

VAN 最大的特点是通过计算机服务网络，可使不同企业、不同网络系统相互连接，从而使不同形式的数据交换成为可能。由于 VAN 实现了不同系统的对接和不同格式信息的交换，为无数的使用者提供了交换数据的服务，创造了附加价值，因而被称作"增值网络"。

3. Though E-commerce is on Business-to-Business (B2B), Business-to-consumer (B2C), it is being increasingly realized that the B2B holds the most potential.
though 引导让步状语从句。
本句可译为：虽然电子商务以 B2B 和 B2C 交易为主，但是人们越来越意识到 B2B 更具发展前景。

Learn and Practice

1. Mark the following statements with T (true) or F (false) according to the text.

（1）E-commerce can be divided into three main categories：B2B, B2C, and C2C. （   ）
（2）Transaction between the wholesaler and retailer belongs to B2C mode. （   ）
（3）Online trading is a kind of E-commerce. （   ）

4) B2B is the main transaction of E-commerce. （　　）

2. Translate the following phrases into Chinese or English.

1) 电子商务。

2) file taxes.

3) E-commerce model.

4) procurement process.

3. Translate the following passage into Chinese.

E-commerce or electronic commerce is the purchasing, selling, and exchanging of goods and services over computer networks (such as the Internet) through which transactions or terms of sale are performed electronically.

Extensive Reading

Project Management

Project management is the application of knowledge, skills, tools and techniques to project activities to meet project requirements. Project management is accomplished through the application and *integration* of the project management processes of initiating, planning, executing, monitoring and controlling, and closing. [1] The project manager is the person responsible for accomplishing the project objectives.

Managing a project includes:

1) Identifying requirements.

2) Establishing clear and achievable objectives.

3) Balancing the competing demands for quality, scope, time and cost.

4) Adapting the specifications, plans, and approaches to the different concerns and *expectations* of the various *stakeholders*.

It is important to note that many of the processes within project management are interactive because of the existence of, and necessity for, progressive *elaboration* in a project throughout the project's life cycle. [2] That is, as a project management team learns more about a project, the team can then manage to a greater level of detail.

The term "project management" is sometimes used to describe an organizational or managerial approach to the management of projects and some ongoing operations, which can be redefined as projects, that is also referred to as "management by projects". An organization that adopts this approach defines its activities as projects in a way that is consistent with the definition of a project. There has been a tendency in recent years to manage more activities in more application areas using project management. More organizations are using "management by projects". This is not to say that all operations can or should be organized into projects. The *adoption* of "management by projects" is also related to the adoption of an organizational culture that is close to the project manage-

Unit 14
E-commerce

ment culture. [3] Although, an understanding of project management is *critical* to an organization that is using "management by projects", a detailed discussion of the approach itself is outside the scope of this standard.

Words and Expressions

integration [ˌɪntɪˈɡreɪʃn] n. 综合，一体化
expectation [ˌekspekˈteɪʃn] n. 期待，期望
stakeholder [ˈsteɪkhəʊldər] n. 股东；利益相关者
elaboration [ɪˌlæbəˈreɪʃn] n. 精心制作；详尽
adoption [əˈdɒpʃn] n. 采取；收养
critical [ˈkrɪtɪkl] adj. 批评的；决定性的

Special Difficulties

1. Project management is accomplished through the application and integration of the project management processes of initiating, planning, executing, monitoring and controlling, and closing.

本句可译为：项目管理是通过项目管理过程中的启动、计划、执行、监控和结束这些步骤的应用和集成来实现的。

2. It is important to note that many of the processes within project management are interactive because of the existence of, and necessity for, progressive elaboration in a project throughout the project's life cycle.

progressive elaboration 意为"进一步细化"。

本句可译为：需要注意的是，在整个项目生命周期中，由于事项逐步完善情况的存在和必然性，项目管理中很多过程是相互影响的。

3. The adoption of "management by projects" is also related to the adoption of an organizational culture that is close to the project management culture.

management by projects 意为"项目化管理"。

本句可译为："项目化管理"的采用还与采用接近项目管理文化的组织文化有关。

Learn and Practice

1. Mark the following statements with T (true) or F (false) according to the text.

1) Managing a project includes establishing clear and achievable objectives. (   )

2) It is not important to note that many of the processes within project management are interactive. (   )

3) Less organizations are using "management by projects". (   )

2. Give short answers to the questions according to the text.

1) What is project management and how is it accomplished?

2) What are the points included in managing a project?

3) How is the development of "management by projects" in nowadays?

Knowledge Link

科技论文的写作和发表 2

一、科技论文的选题

科技论文选题是指论文要论述的范围或研究方向，通常是研究过程中选定的研究课题。尽管可供研究的课题数不胜数，但并不是任何选题都具有研究价值，可以形成论文。选题既会受到自己学术水平、研究能力的限制，又会受到研究条件的制约。因此，我们应选择那些既能反映自己学术水平和创新能力、又符合客观研究条件的课题。

1. 选题的确定原则

（1）创造性原则　科技论文是对自己学习成果和科学研究的总结，论文的选题要能反映学习和科学研究中取得的成绩，要在前人的研究基础上有所创新。人类社会之所以不断向前发展，关键就在于不断创造新的成果。衡量一篇论文是否具有价值，关键在于其是否具有新的内容或新的研究方法。因此，创新是一篇论文必须遵循的原则。在论文的选题确定阶段，就要注意论文是否具有新意，老调重弹、拾人牙慧的论文往往不会具有创新性，也不会产生好的社会影响。

（2）科学性原则　科学性原则是指论文必须具有科学价值，论文的科学性原则，一方面要求论文能反映社会的现实需要，根据社会现实确定论文选题；另一方面要求论文能反映科学研究的最新进展，根据科学研究的实际情况确定论文选题。也就是说，论文要符合科学和社会发展的规律，仅有创新性而不具备科学性的论文，同样也会毫无价值。

（3）可行性原则　由于论文选题往往受到主观和客观条件的限制，因此，论文的选题确定还要遵循可行性的原则。有些论文选题虽然非常好，价值非常大，但由于作者自身条件或研究条件等客观条件的限制，即使选择了这一选题，最后也无法完成。在实际确定选题的过程中，必须实事求是地从主客观实际出发，恰当地把握选题的范围大小、难易程度。

（4）选题的时间　选题要尽可能早地确定，尤其是对大学生写作毕业论文而言。因为毕业论文反映了大学期间的学习成果，是对大学生活的总结。早做准备，时间充分，能够比较从容地从事调查和研究。搜集资料的过程不是一朝一夕就能完成的，一些本校没有的资料还要到校外进行查找，更要花费不少的时间。当然，也不是越早确定选题越好，要根据专业课的学习情况而定。过早的话，由于缺少必要的专业基础知识，很难发现和评价选题的优劣；但过晚的话，来不及仔细地调查研究和认真思考，容易草率结题。

（5）选题的大小　一般而言，毕业论文选题范围不宜过大、涉及面不宜太宽。因为范围过大，不但时间不允许，而且很可能缺乏这种功力。即使勉强写下来了，也只能如蜻蜓点水一般，难以保证质量。论文的选题小一些、专一些，既容易完成，也容易写好。一些人认为，只有写大题目才算写论文，这是一种极大的误解。当然，选题也不能太小，太小的选题，搜集资料、阐述都不容易，也达不到锻炼和提高的目的。

（6）选题的难易　选题的难易程度要合适，既不可过难又不可过易。对初写学术论文的作者来说，选择过高难度的课题，不仅达不到提高研究能力的目的，反而会因写作难度大，挫伤写作的积极性。但过易的题目，又体现不出自己的知识水准和创造性，同样不利于学术水平的提高。

2. 选题的确定方式

常见的论文选题确定方式有主动选题、被动选题、主动与被动相结合的选题三种。

主动选题是论文作者自己提出的选题。论文作者在自己的学习和工作中，根据自己的兴趣、爱好，就某些问题进行认真探讨，发现某些问题前人没有探讨过，或某些问题前人研讨还不够深入，就可在此基础上形成论文的选题。尤其是在学习或研究过程中，发现自己不清楚的问题，或发现别人对某些问题具有错误的认识，更能形成好的选题。

被动选题是指别人提出的论文选题。由于提出选题的人往往具有较高的知识水准，能够把握学术发展的脉搏，因此，这些选题往往具有较大的价值。被动选题如果与自己的知识结构、兴趣爱好相吻合，往往容易完成，而且容易写好。但如果与自己的知识结构、兴趣爱好等不相吻合，完成的难度可能较大。

主动与被动相结合的选题是指论文作者提出某选题后，他人认为该选题虽好，但存在某些问题，因此加以更改而形成的更好的选题；或是他人提出某选题后，论文作者根据自己的实际情况进行更改而形成的更符合自己情况的选题。主动与被动相结合的选题往往更加适应作者的实际情况，具有更大的可行性，或更具有科学性和创新性。

3. 选题的调研

要确定一个好的论文选题必须进行认真调研。尤其是进行主动选题时，更要做好论文选题的调研工作，如该选题的研究现状和发展情况，已经取得的成绩和存在的不足，有代表性的论文和著作等。如果不做充分调研的话，自己辛辛苦苦写出的论文，别人早就做过，或已经被别人证明不再具备科学研究的价值，那就白白浪费了时间和精力。如果发现该选题前人已做了大量的研究工作，但至今仍是一个有争议的问题，而自己又有不同于前人的看法，或虽与某一家之说相同，但自己挖掘出了新的论据资料，那么，这个选题仍然具有研究的价值。确定论文选题后就要开始进行资料的收集，好的选题需要充足的材料支撑，科技文献检索与利用在前面的章节中已做了详细介绍，这里就不再赘述。

二、科技论文的写作

1. 编写提纲

编写提纲是作者动笔行文前的必要准备，是作者构思谋篇的具体体现。有了一个好的提纲，就能纲举目张，掌握全篇论文的基本骨架，使论文的结构完整统一；能分清层次，明确重点，周密地谋篇布局，使总论点和分论点有机地统一起来；也就能够按照各部分的要求安排、组织、利用资料，决定取舍，最大限度地发挥资料的作用。

论文提纲可分为简单提纲和详细提纲两种。简单提纲是高度概括的，只提示论文的要点，如何展开则不涉及。这种提纲虽然简单，但由于它是经过深思熟虑构成的，因此写作时能顺利进行。如果没有这种准备，边想边写，则很难顺利地写下去。

详细提纲，是把论文的主要论点和展开思路较为详细地列出来。如果在写作之前准备了详细提纲,那么,写作时就能更加顺利。

2. 论文写作

为便于论文所报道的科学技术研究成果这一信息的收集、储存、处理、加工、检索、利用、交流和传播,国家标准对科技论文的撰写和编排格式做了规定。尽管各篇论文的内容千差万别,不同作者的写作风格各不相同,但格式完全可以统一。对于一篇科技论文,应先写什么,后写什么,各部分要写什么内容,以及表达中有些什么要求,编排上应符合哪些规定,都有章可循。但是,论文的主题如何确立,论据如何选取,论证如何进行,结构如何安排,节、段如何划分,层次标题如何拟定等,则需要论文作者和刊物编者根据具体的研究对象、研究目的和方法以及论文内容进行处理,即根据实际情况来处理。可以说,只要按照科技论文的撰写和编排格式去进行创造性的写作和编辑,就能写出既符合规定格式要求又各自具有独立的主题思想、表达手法、写作风格和编排特色的高质量论文。

论文是用来表述研究成果的,只有选用最贴切、最恰当的词汇,组织既简练又符合语法规范的句子,才能够把研究者的研究成果准确、鲜明、充分地表述出来。因此,在论文的具体写作过程中,要求做到语言准确、精练,行文流畅,这是学术论文的基本要求。此外,在论文写作中还要注意提炼中心句。为了方便读者对论文的阅读和理解,论文中的每个段落都应该有个中心句。该句能准确反映该段的主旨和核心内容,使读者阅读该句就能对该段的内容有大概的了解。中心句一般位于段落的开头或结尾。

3. 资料引用

在撰写学术论文的过程中,往往要引用其他作者的有关研究成果。文献的引用一般有两种方式:一种是直接引用,即将原文直接照抄,将抄录的原文用引号标注,或用其他字体另起一段排列;一种是间接引用,通过自己的语言,表述原文的内容。但不管是直接引用还是间接引用,都需要标明原始文献的出处。有些人认为,参考别人的观点而未引用别人的材料,就不需要标明出处。这种观点其实是错误的,无论是引用他人的观点还是他人的材料,都需要明确地标明出处。引用他人的研究成果,要求准确地引用原文,切不可断章取义。对于一些未公开发表的文献,一般不引为宜。一方面因为这些文献尚未得到社会的公开承认;另一方面,如果首先引用的话,披露资料中的一些重要材料,会对作者的首创权产生一定的负面影响。

引用文献的注释方式主要有三种:一种是夹注,也称"文中注",即将引用文献的出处直接标注在引文的后面。报纸上的引文大多采用这种方式。这种方式的好处是一目了然,读者读完引文后就知道出处;缺点是容易破坏原文的结构和阅读节奏,且不易了解整篇文章引用的所有引文。第二种是脚注,也称"页下注",注在一页文章的最下面,即通常在页脚位置标出引文的出处。页下注的好处是清晰、明了,能知道该页文章中所引用的全部引文;缺点是引文字数不能太多,否则版面的排列容易受到影响。第三种是篇末注,也称"文后注",在一篇文章或一章、一节内容的最后标出所有引文的出处。篇末注的好处是能清晰地展示整篇文章(或一章、一节的内容)引用的文献出处,同时可以不受字数的限制;缺点是读者如果看完引文后要立即翻阅出处的话,不够方便。期刊、图书大多采用页下注或篇后注的方式。无论采用哪种注释方式,都要准确标出原文的作者、题名、出处、时间、原文中的位置,同时还要采用标准的著录法。

4. 论文修改

好论文是写出来的，同时也是修改出来的。论文写作是一种写作能力的锻炼和综合能力的训练，要提高写作能力，既要多写，更要多改。在修改论文之前，首先要仔细阅读，只有这样才能发现文章存在的问题，也可请领域内相关专家或同行阅读，以发现论文中存在的问题。

三、科技论文的发表

论文只有正式发表才能产生影响，并接受社会的评价。因此，将修改后的论文进行投稿是论文产生社会影响的关键。论文投稿的关键是选择恰当的期刊，有些人过高地估计了自己的论文水平，一开始就向一些权威期刊投稿，结果石沉大海，挫伤了投稿的积极性。有些人则不知道向何处投稿，将文章投向了与论文主题不相关的期刊，同样影响了论文的刊用。因此，不仅仅要会写论文，还要会投稿。

对初次投稿者而言，选择合适的期刊非常重要。初次投稿可以请老师或者行业专家推荐，当然也可以自己利用检索工具选择合适的期刊进行投稿。五大机构（中科院、中信所、社科院、北大、南大）核心期刊一览表在网上均可查到；各学科期刊、会议论文集、标准、学位论文等登陆中国知网也都可查询了解。当然，了解期刊的最好方式是到一些大中型图书馆、大学资料室等处去阅读这些期刊，这样就能对这些期刊有更充分的了解。

初次投稿如果没有人推荐和指导，投稿者往往很难合理选择要投的刊物。如果投稿者的研究水平不够，初次投稿往往也不容易被采用。因此，建议初次投稿先选择非核心期刊，这是因为非核心期刊的要求相对没有那么高。待自己水平提高后，再向核心期刊投稿。如果第一次投稿未被采用，可再选择一些刊物进行尝试。当然，投稿时也要遵循一定的道德规范，不能同时一稿多投，因为如果稿件同时被多个期刊采用的话，既浪费了其他刊物宝贵的版面，也浪费了别人的时间和劳动。按照我国的《著作权法》以及各报刊的一般约定，如果投出的文章三个月内没有收到被采用的消息，作者可以再投稿。如果方便的话，可以向投稿的编辑部询问一下文章的处理结果，以免产生不必要的误会。因为有些刊物发现作者有一稿多投的行为后，会不再刊登该作者的其他稿件。

Dialogue Exercise

Background: In the emerging global economy, E-commerce has increasingly become a necessary component of our daily life. Li Lei and Han Meimei are talking about the E-commerce.

Han Meimei: E-commerce is very convenient and efficient. But I am just wondering when and how E-commerce came into being.

Li Lei: Oh, it is a big question. I think I can answer you because I have just read an article about the development of E-commerce.

Han Meimei: So, I am asking the right person?

Li Lei: Yes. The Internet was born in 1969, with time passing by, the number of host comput-

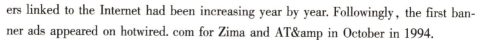

ers linked to the Internet had been increasing year by year. Followingly, the first banner ads appeared on hotwired. com for Zima and AT&amp in October in 1994.

Han Meimei: Do you mean that the Internet was used for commercial purposes only during the 1990s?

Li Lei: Right. Since the first banner ads, companies have used the Internet well beyond more promotion. Nowadays, goods and services can be bought and paid online; global auctions can be conducted.

Han Meimei: I see. It's the Internet that has facilitated the buying and selling of goods and services.

> **Famous Quotes**
>
> **True mastery of any skill takes a lifetime.**
> 对任何技能的掌握都需要一生的刻苦操练。

Appendixes

Appendix A 词汇索引

A

a bank of 一系列；一排 Unit 2
ABS 防抱死制动系统 Unit 12
academic [ˌækəˈdemɪk] adj. 学院的，大学的，学会的 Unit 4
accessory [əkˈsesəri] n. 配件，附件 Unit 3
accommodate [əˈkɒmədeɪt] v. 容纳，使适应；调解 Unit 2
accumulate [əˈkjuːmjəleɪt] v. 积累，积聚，堆积 Unit 3
accumulation [əˌkjuːmjəˈleɪʃn] n. 积累；堆积物；累积量 Unit 1
accuracy [ˈækjərəsi] n. 精确（性），准确（性） Unit 8
accurately [ˈækjərətli] adv. 正确无误地，准确地 Unit 9
acid fog 酸雾 Unit 12
adoption [əˈdɒpʃn] n. 采取；收养 Unit 14
aerospace [ˈeərəʊspeɪs] n. 航空航天工业 Unit 6
airbag [ˈeəbæɡ] n. 安全气囊 Unit 12
algorithm [ˈælɡərɪðəm] n. 算法，运算法则 Unit 3
ambiguous [æmˈbɪɡjuəs] adj. 含糊的，不明确的 Unit 9
analysis [əˈnæləsɪs] n. 分析；分解；梗概 Unit 1
angles [ˈæŋɡlz] n. 角；角度 Unit 2
angular [ˈæŋɡjələr] adj. 有角的，用角测量的 Unit 5
appealing [əˈpiːlɪŋ] adj. 吸引人的，令人心动的 Unit 9
application [ˌæplɪˈkeɪʃn] n. 应用；申请；用途 Unit 13
approach [əˈprəʊtʃ] n. 方法；接近 Unit 6
appropriately [əˈprəʊpriətli] adv. 适当地 Unit 1
approximate [əˈprɒksɪmət] adj. 大概的；极相似的 Unit 1
apron [ˈeɪprən] n. 溜板 Unit 6
artificial [ˌɑːtɪˈfɪʃl] adj. 人造的；虚假的；非原产地的。 Unit 10
Artificial Intelligence（AI）人工智能 Unit 1
artificial satellites 人造卫星 Unit 10

ASP (analog signal processing) 模拟信号处理 Unit 11
aspiration [ˌæspəˈreɪʃn] n. 强烈的愿望；吸气 Unit 2
assertive [əˈsɜːtɪv] adj. 观点明确的；坚决主张的 Unit 9
astronaut [ˈæstrənɔːt] n. 宇航员，太空人 Unit 10
astronomical [ˌæstrəˈnɒmɪkl] adj. 天文学的；极大的 Unit 10
atmosphere [ˈætməsfɪə(r)] n. 大气；风格；气氛 Unit 10
attain [əˈteɪn] v. 达到，获得 Unit 8
authentication [ɔːˌθentɪˈkeɪʃn] n. 鉴定，证明；身份验证 Unit 13
Auto-ID technology 自动识别技术 Unit 9
automated driving 自动驾驶 Unit 12
autonomy [ɔːˈtɒnəmi] n. 自治；自主权；人身自由 Unit 1

B

beam [biːm] n. 梁；光线；（电波的）波束 Unit 4
benefit [ˈbenɪfɪt] n. 利益，好处；救济金 Unit 3
biometric [ˌbaɪəʊˈmetrɪk] adj. 生物统计学的 Unit 9
bioinformatics [ˌbiːəʊɪnˈfɔːmætɪks] n. 生物信息学 Unit 9
blend [blend] v. 混合；掺杂；结合 Unit 9
boring [ˈbɔːrɪŋ] n. 镗削 Unit 6
bottleneck [ˈbɒtlnek] n. 瓶颈；瓶颈路段 Unit 8
brake [breɪk] n. 制动器；阻碍 Unit 10

C

cache memory 高速缓冲存储器，高速缓存 Unit 11
campaign [kæmˈpeɪn] n. 运动；战役 Unit 5
capability [ˌkeɪpəˈbɪləti] n. 能力；性能；容量 Unit 13
capacity [kəˈpæsəti] n. 容量；才能；性能 Unit 1
capital and labor equipments 资金和劳动力需求 Unit 6
carriage [ˈkærɪdʒ] n. 刀架 Unit 6
cast [kɑːst] vt. 铸造 Unit 5
catalog [ˈkætəlɒɡ] n. 目录；产品样本 Unit 14
characterization [ˌkærəktəraɪˈzeɪʃn] n. 特性描述；刻画，塑造 Unit 1
characterize [ˈkærəktəraɪz] vt. 表现……的特色，具有……的特征 Unit 2
CIMS (computer-intergrated manufacturing system) 计算机集成制造系统 Unit 2
code [kəʊd] v. 将……译成电码，编码；加密 Unit 10
cognitive science 认知科学 Unit 1
columniform [kəˈlʌmnɪfɔːm] adj. 圆柱形的 Unit 10
combination [ˌkɒmbɪˈneɪʃn] n. 结合；组合；联合 Unit 13
competitive [kəmˈpetətɪv] adj. 竞争的，比赛的 Unit 2

complex [ˈkɒmpleks] adj. 复杂的；合成的 Unit 12
component [kəmˈpəʊnənt] n. 构成；组件 Unit 6
compression [kəmˈpreʃən] n. 压缩；浓缩 Unit 11
comprise [kəmˈpraɪz] v. 包含；由……组成 Unit 13
concentrate [ˌkɒnsnˈtreɪt] v. 专心；关注 Unit 10
confidentiality [ˌkɒnfɪˌdenʃiˈæləti] n. 机密性 Unit 13
consistent [kənˈsɪstənt] adj. 一致的，连续的 Unit 2
consolidate [kənˈsɒlɪdeɪt] v. 巩固，使固定；联合 Unit 3
constitute [ˈkɒnstɪtjuːt] v. 构成；组成 Unit 8
consume [kənˈsjuːm] v. 消耗；消费 Unit 8
contour [ˈkɒntʊər] n. 外形，轮廓 Unit 5
contouring [kɔnˈtʊərɪŋ] n. 轮廓，造型 Unit 6
control unit 控制器，控制部件 Unit 11
conveyor [kənˈveɪər] n. 传输；输送机 Unit 6
corrupted [kəˈrʌptɪd] v. 败坏；腐化 Unit 13
critical [ˈkrɪtɪkl] adj. 批评的；决定性的 Unit 14
cross motion 横向运动 Unit 6
crucial [ˈkruːʃl] adj. 关键性的，极其显要的 Unit 1
cruise control 巡航控制 Unit 12
cubicle [ˈkjuːbɪkl] n. 小卧室，斗室 Unit 2
curved surface 曲面 Unit 6
Customer Relationship Management 客户关系管理 Unit 3

D

decipher [dɪˈsaɪfər] v. 译解，解释 Unit 11
definition [ˌdefɪˈnɪʃn] n. 定义；规定 Unit 14
detection [dɪˈtekʃn] n. 侦查；检查 Unit 8
detract [dɪˈtrækt] v. 贬低；减损 Unit 8
device [dɪˈvaɪs] n. 装置；策略；设备 Unit 13
dexterity [dekˈsterəti] n. 灵巧，熟练 Unit 5
diagnostics [ˌdaɪəgˈnɒstɪks] n. 诊断学 Unit 7
digital model 数字模型 Unit 4
dimension [daɪˈmenʃn] n. 尺寸；范围；维度 Unit 1
disclosure [dɪsˈkləʊʒə(r)] n. 公开；泄露 Unit 13
distinguish [dɪˈstɪŋgwɪʃ] v. 区分，辨别 Unit 11
docking [ˈdɒkɪŋ] v. 对接；减少 Unit 10
domain [dəʊˈmeɪn] n. 范围；领域 Unit 7
drastically [ˈdrɑːstɪklɪ] adv. 彻底地，激烈地 Unit 4
drilling [ˈdrɪlɪŋ] n. 钻削 Unit 6

DSP (digital signal processing) 数字信号处理 Unit 11

E

earth orbit 环地轨道 Unit 10
efficiency [ɪˈfɪʃnsi] n. 效率；能力 Unit 4
efficiently [ɪˈfɪʃntlɪ] adv. 有效地；效率高地 Unit 13
elaboration [ɪˌlæbəˈreɪʃn] n. 精心制作；详尽 Unit 14
electrical discharge forming machine 电火花成型机床 Unit 6
electromagnetic [ɪˌlektrəʊmæɡˈnetɪk] adj. 电磁的 Unit 9
Electronic Commerce 电子商务 Unit 14
electronically [ɪˌlekˈtrɒnɪkli] adv. 电子地 Unit 14
elicit [iˈlɪsɪt] v. 探出；引出 Unit 9
emphasis [ˈemfəsɪs] n. 强调，突出 Unit 4
encryption [ɪnˈkrɪpʃn] n. 编码，加密 Unit 13
end effector 终端执行器 Unit 5
Enterprise Resource Planning 企业资源管理 Unit 3
essential [ɪˈsenʃl] adj. 必要的；本质的；精华的 Unit 1
estimation [ˌestɪˈmeɪʃn] n. 估算；评估 Unit 6
evaluate [ɪˈvæljueɪt] v. 评价；求……的值（或数） Unit 4
excel in 在...方面胜出 Unit 1
executive [ɪɡˈzekjətɪv] adj. 行政的；经营的；执行的 Unit 3
expectation [ˌekspekˈteɪʃn] n. 期待，期望 Unit 14
expert system 专家系统 Unit 1
expose [ɪkˈspəʊz] v. 揭露，揭发；使暴露 Unit 13
extrude [ɪkˈstruːd] v. （被）挤压出；喷出 Unit 4

F

facilitate [fəˈsɪlɪteɪt] v. 促进；帮助；使容易 Unit 13
FFT (Fast Fourier Transform Algorithm) 快速傅氏变换算法 Unit 11
financial [faɪˈnænʃl] adj. 财政的，财物的 Unit 13
flexibility [ˌfleksəˈbɪlətɪ] n. 柔韧性，灵活性 Unit 11
flexible [ˈfleksəbl] adj. 灵活的；柔性的 Unit 6
FMS (flexible manufacturing system) 柔性制造系统 Unit 2
forefront [ˈfɔːfrʌnt] n. 最前方；活动中心 Unit 10
forge [fɔːdʒ] vt. 锻造；伪造 Unit 5
formation [fɔːˈmeɪʃn] n. 形成；结构 Unit 3
fossil fuel 化石燃料，矿物燃料 Unit 12
frames [freɪmz] n. 框架；边框 Unit 4
fundamental [ˌfʌndəˈmentl] n. 原理；基本，基础 Unit 1

fusion ['fjuːʒn] n. 融合；合并 Unit 8

G

gage [geidʒ] vt. 计量，度量 Unit 5
gantry ['gæntri] n. 构台，桶架，台架 Unit 5
gear cutting 齿轮加工 Unit 6
GPS 全球定位系统 Unit 12
grinding machine 磨床 Unit 6
guarantee [ˌgærən'tiː] v. 保证，担保 Unit 8

H

hobbyist ['hɒbiːɪst] n. 爱好者 Unit 4
holo-deck 全息驾驶舱 Unit 2
horizontal milling machine 卧式铣床 Unit 6
hybrid ['haɪbrɪd] adj. 混合的；杂交的 Unit 12
hyphenated ['haɪfəneɪtɪd] adj. 带有连字符的 Unit 9

I

identification [aɪˌdentɪfɪ'keɪʃn] n. 鉴定；识别；验明 Unit 9
illusion [ɪ'luːʒn] n. 错觉，幻觉 Unit 2
immerse [ɪ'mɜːs] vt. 浸没；陷入 Unit 2
implementing ['impliməntɪŋ] v. 实现；执行；使生效 Unit 4
incoming ['ɪnkʌmɪŋ] adj. 进来的，回来的；新到的 Unit 8
incorporated [ɪn'kɔːpəreɪtɪd] adj. 股份有限的，组成公司的 Unit 4
incredibly [ɪn'kredəbli] adv. 难以置信地；极其 Unit 11
individual [ˌɪndɪ'vɪdʒuəl] adj. 个人的；独立的 Unit 9
infrared rays 红外线 Unit 10
ingenious [ɪn'dʒiːniəs] adj. 设计独特的，巧妙的；精致的 Unit 12
innocuous [ɪ'nɒkjuəs] adj. 无害的；平淡无味的 Unit 13
inspired [ɪn'spaɪəd] v. 鼓舞；激励 Unit 4
instrument ['ɪnstrəmənt] n. 仪器；工具；乐器。 Unit 10
integrated ['ɪntɪgreɪtɪd] adj. 完整的；整体的；结合的 Unit 4
integration [ˌɪntɪ'greɪʃn] n. 综合，一体化 Unit 14
intelligent [ɪn'telɪdʒənt] adj. 聪明的；智能的 Unit 1
interaction [ˌɪntər'ækʃn] n. 合作；互相影响；互动 Unit 2
interchangeable [ˌɪntə'tʃeɪndʒəbl] adj. 可交换的；可交替的 Unit 6
intervention [ˌɪntə'venʃn] n. 介入，干预 Unit 3
involvement [ɪn'vɒlvmənt] n. 牵连；参与；加入 Unit 9
IOT 物联网 Unit 3

K

keyboard [ˈkiːbɔːd] n. 键盘 Unit 11
knowledge representation　知识表示 Unit 1

L

laudatory [ˈlɔːdətəri] adj. 表扬的，赞扬的 Unit 12
lead-times 交付周期；更换模具的时间 Unit 4
leap [liːp] n. &v. 跳跃，飞越 Unit 9
license [ˈlaɪsns] n. 许可证，执照；特许 Unit 14

M

M2M 机器通信 Unit 3
machining center 加工中心 Unit 6
magnetic [mæɡˈnetɪk] adj. 有磁性的；有吸引力的 Unit 9
maintain [meɪnˈteɪn] vt. 保持；保养 Unit 2
maneuver [məˈnuːvə] v. 移动；操纵 Unit 10
manipulate [məˈnɪpjuleɪt] vt. 操作，处理 Unit 2
manipulator [məˈnɪpjuleɪtər] n. 操纵者；操纵器 Unit 5
manual inspection methods　手工检测方式 Unit 8
marvelous [ˈmɑːvələs] adj. 不可思议的；非凡的 Unit 1
mathematical [ˌmæθəˈmætɪkl] adj. 数学的，数学上的；精确的 Unit 3
megahertz [ˈmeɡəhɜːts] n. 兆赫（MHz）Unit 11
methodology [ˌmeθəˈdɒlədʒi] n. 方法学，方法论 Unit 4
MICR（magnetic ink character recognition）磁墨水字符识别 Unit 9
microprocessor [ˌmaɪkrəʊˈprəʊsesər] n. 微处理器 Unit 11
milestone [ˈmaɪlstəʊn] n. 里程碑；划时代事件 Unit 9
military [ˈmɪlətri] adj. 军事的，军用的 Unit 8
milling [ˈmɪlɪŋ] n. 铣削 Unit 6
minimize [ˈmɪnɪmaɪz] vt. 把……减至最低数量（程度）Unit 5
modify [ˈmɒdɪfaɪ] v. 修改，更改 Unit 4
modular production system 模块化生产系统 Unit 5
modulation [ˌmɒdjuˈleɪʃn] n. 调节；调制，调整 Unit 12
module [ˈmɒdjuːl] n. 舱；模块 Unit 10
molecular [məˈlekjələr] adj. 分子的，由分子组成的 Unit 1
monitor [ˈmɒnɪtə(r)] v. 监控，监听 Unit 7
monotonous [məˈnɒtənəs] adj. 枯燥的；（声音）单调的 Unit 8
mouse [maʊs] n. 鼠标，鼠标器　mice [复数] Unit 11
MSP（mixed signal processing）混合信号处理 Unit 11

N

nagging [ˈnægɪŋ] adj. 挑剔的，纠缠的 Unit 12
nascent [ˈnæsnt] adj. 初期的；初生的 Unit 3
necessitate [nəˈsesɪteɪt] v. 需要，迫使 Unit 12
negotiable [nɪˈɡəʊʃiəbl] adj. 可谈判的；可协商的 Unit 14

O

obligation [ˌɒblɪˈɡeɪʃn] n. 义务，责任；契约 Unit 13
operational [ˌɒpəˈreɪʃənl] adj. 操作的；运作的 Unit 3
opposite [ˈɒpəzɪt] adj. 相对的，对立的 Unit 12
optimize [ˈɒptɪmaɪz] adj. 最佳化的，（使）最优化的 Unit 4
orbital [ˈɔːbɪtl] adj. 轨道的 Unit 10
organism [ˈɔːɡənɪzəm] n. 有机体；生物体 Unit 9
oriented [ˈɔːrientɪd] adj. 导向的；定向的 Unit 1
oscillator [ˈɒsɪleɪtər] n. 振荡器 Unit 11
override [ˌəʊvəˈraɪd] vt. 推翻；优先于；覆盖 Unit 7
oversee [ˌəʊvəˈsiː] v. 俯瞰；监视 Unit 12
ozone [ˈəʊzəʊn] n. 臭氧；新鲜的空气 Unit 12

P

pallet [ˈpælət] n. 托盘，托板；平台 Unit 6
participant [pɑːˈtɪsɪpənt] n. 参加者，参与者 Unit 2
patent [ˈpætnt] n. 专利；专利品 Unit 4
penetrate [ˈpenətreɪt] v. 穿透；渗入；洞悉 Unit 10
perceive [pəˈsiːv] v. 感觉；理解为 Unit 4
performance [pəˈfɔːməns] n. 性能；表演；表现 Unit 3
phenomenon [fəˈnɒmɪnən] n. 现象；事件 Unit 4
photochemical smog 光化烟雾 Unit 12
pluralistic [ˌplʊərəˈlɪstɪk] adj. 兼职的；多元化的 Unit 4
polarized [ˈpəʊləraɪzd] vi. 使极化；使偏震 Unit 2
power supply 电力供应 Unit 5
precise [prɪˈsaɪs] adj. 精确的，精密的 Unit 8
preference [ˈprefrəns] n. 偏爱；优先权 Unit 2
primarily [praɪˈmerəli] adv. 根本地；首要地；主要地 Unit 9
primary [ˈpraɪməri] adj. 首要的；原始的；原生的 Unit 1
printer [ˈprɪntər] n. 打印机 Unit 11
prioritize [praɪˈɒrətaɪz] vt. 按重要性排列；优先处理 Unit 4
probe [prəʊb] n. 探针；探索；探测仪 Unit 10

procurement [prəˈkjuəmənt] n. 获得，取得；采购 Unit 14
production [prəˈdʌkʃn] n. 生产，制作 Unit 5
progressive [prəˈgresɪv] adj. 进步的；先进的 Unit 3
propagation [ˌprɒpəˈgeɪʃn] n. 传播，传输；蔓延 Unit 9
propellant [prəˈpelənt] adj. 推进的 n. 推进物 Unit 10
propulsion [prəˈpʌlʃn] n. 推进，推动力，驱动 Unit 12
purchase [ˈpɜːtʃəs] v. 购买；换取 Unit 14

Q

quality control 质量监控 Unit 8
quantity [ˈkwɒntəti] n. 数量；大批 Unit 2

R

ream [riːm] v. 扩孔 Unit 5
reaming [ˈriːmɪŋ] n. 铰削 Unit 6
rechargeable battery pack 可再充电电池组 Unit 12
recognition [ˌrekəgˈnɪʃn] n. 认识，识别；承认，认可 Unit 9
reconnect [ˌriːkəˈnekt] vt. 再供应，再接通 Unit 7
refrigerator [rɪˈfrɪdʒəreɪtə(r)] n. 冰箱 Unit 3
relative movement 相对运动 Unit 5
remodel [ˌriːˈmɒdl] v. 改造；改变；改型 Unit 3
replica [ˈreplɪkə] n. 复制品 Unit 4
repudiation [rɪˌpjuːdɪˈeɪʃn] n. 拒绝；否认 Unit 13
resin [ˈrezɪn] n. 树脂；松香 Unit 4
responsible [rɪˈspɒnsəbl] adj. 有责任的，负责的 Unit 12
retailer [ˈriːteɪlər] n. 零售商，零售店 Unit 14
retrieval [rɪˈtriːvl] n. 恢复；取回 Unit 6
RFID (Radio Frequency Identification) 射频识别技术 Unit 3
rigorous [ˈrɪgərəs] adj. 严格的，严厉的 Unit 12
rival [ˈraɪvl] v. 竞争；匹敌 Unit 12
robotics [rəʊˈbɒtɪks] n. 机器人技术 Unit 5
rod [rɒd] n. 杆，拉杆 Unit 5
rotary [ˈrəʊtəri] adj. 旋转的，转动的；轮流的 Unit 12

S

satellite [ˈsætəlaɪt] n. 卫星；人造卫星；Unit 13
scheme [skiːm] n. 计划；体系；阴谋 Unit 8
scope [skəʊp] n. 范围；余地 Unit 10
scrap [skræp] v. 废弃，报废 Unit 8

secure [sɪˈkjʊə (r)] v. 获得安全，变得安全 Unit 5
security [sɪˈkjʊərəti] adj. 安全的；保密的 Unit 3
shop floor 车间，厂房 Unit 2
significant [sɪɡˈnɪfɪkənt] adj. 重要的；显著的；意味深长的 Unit 1
simulate [ˈsɪmjuleɪt] vt. 假装；模仿；模拟 Unit 2
software [ˈsɒftweə (r)] n. 软件 Unit 13
solar cell 太阳能电池 Unit 10
sonar [ˈsəʊnɑːr] n. 声呐装置 Unit 11
spaceship [ˈspeɪsʃɪp] n. 宇宙飞船 Unit 10
spindle [ˈspɪndl] n. 主轴 Unit 6
stakeholder [ˈsteɪkhəʊldər] n. 股东；利益相关者 Unit 14
Star Trek 星际迷航（美国科幻影视作品）Unit 2
starter [ˈstɑːtə] n. （发动机的）起动装置；参赛人 Unit 7
statistical procedures 统计程序 Unit 8
stereo [ˈsteriəʊ] n. 立体声；立体声音响（器材）Unit 3
subarea [ˌsʌbˈeərɪə] n. 分区；分支 Unit 1
suborbital [[sʌˈbɔːbɪtl]] adj. 不满轨道一圈的，亚轨道的 Unit 10
substantial [səbˈstænʃl] adj. 充实的；实质的 Unit 2
superfluous [suːˈpɜːfluəs] adj. 过多的，多余的；奢侈的 Unit 14
surveillance [sɜːˈveɪləns] n. 监视，监督 Unit 8
survey [ˈsɜːveɪ] v. 调查；勘测 Unit 8
synergetic [ˌsɪnəˈdʒetɪk] adj. 协同的；协作的 Unit 7

T

tag [tæɡ] vt. 标签；加标签于 Unit 3
tap [tæp] v. 切割 Unit 5
tele-immersion 远程沉浸，远程参与 Unit 2
termination [ˌtɜːmɪˈneɪʃn] n. 终止，终结 Unit 2
thermostat [ˈθɜːməstæt] n. 恒温（调节）器 Unit 1
to date 到目前为止，迄今 Unit 4
tool magazine 刀库 Unit 6
toolholder [ˈtuːlhəʊldər] n. 刀柄 Unit 5
torque [tɔːk] n. （尤指机器的）扭转力，转（力）矩 Unit 5
track [træk] v. 跟踪；检测；追踪 Unit 9
transducer [trænzˈdjuːsər] n. 传感器；变频器 Unit 3
transfer machine 自动生产线 Unit 5
transmission [trænzˈmɪʃn] n. 传动装置；传递；传送 Unit 13
trillion [ˈtrɪljən] n. 万亿，兆 Unit 14
turbine [ˈtɜːbaɪn] n. 涡轮机；汽轮机；叶轮机 Unit 12

turning [ˈtɜːnɪŋ] n. 车削 Unit 6
turning machine 车床 Unit 6

U

ultra-large-scale integrated circuit 超大规模集成电路 Unit 11
ultra-violet light 紫外线 Unit 10
unauthorized [ʌnˈɔːθəraɪzd] adj. 未经授权的；未获允许的 Unit 13
unique [juˈniːk] adj. 唯一的，仅有的 Unit 11
unparalleled [ʌnˈpærəleld] adj. 无比的，无双的 Unit 4

V

vehicle [ˈviːəkl] n. 交通工具；传播媒介 Unit 10
vertical milling machine 立式铣床 Unit 6
vicinity [vəˈsɪnəti] n. 附近；附近地区 Unit 8
video conferencing 视频会议 Unit 2
viewpoint [ˈvjuːpɔint] n. 观点，意见；角度 Unit 7
virtual [ˈvɜːtʃuəl] adj. 实质上的；虚拟的 Unit 2
VLSI（very large scale integrated）超大规模集成电路 Unit 1

W

warranty [ˈwɒrənti] n. 理由，根据；授权 Unit 12
wholesaler [ˈhəʊleɪlər] n. 批发商 Unit 14
wire-cut machine tool 线切割机床 Unit 6
witness [ˈwɪtnəs] v. 作证；表示 Unit 4
workplace reposition 工件再定位 Unit 5

X

X-ray X 射线 Unit 10

#

3D picture 三维图片 Unit 12
3D printing　3D 打印 Unit 4

Appendix B 习题答案

Unit 1

Reading Material
1. F T T
2. 略
3. 参见原文。

Extensive Reading
1. F F T
2. B A C C B
3. 参见译文。

Unit 2

Reading Material
1. F F T F T
2. 参见译文和原文。
3. 参见译文。

Extensive Reading
1. F F T T F
2. 参见译文和原文。
3. 参见译文。

Unit 3

Reading Material
1. F T T
2. A C B C
3. 参见译文。

Extensive Reading
1. F F T T
2. B C B
3. 参见译文。

Unit 4

Reading Material
1. F F F

科技英语

  2. A A A A A

  3. 参见译文。

Extensive Reading

  1. T F F

  2. 参见译文。

Unit 5

Reading Material

  1. T F F T T

  2. C B D

  3. 参见原文和译文。

Extensive Reading

  1. T F T T F

  2. 参见原文和译文。

Unit 6

Reading Material

  1. T F T

  2. 参见译文和原文。

  3. 参见译文。

Extensive Reading

  1. F T F F T F

  2. A B B A C

  3. 参见原文和译文。

Unit 7

Reading Material

  1. F F F T

  2. C B B C

  3. 参见原文和译文。

Extensive Reading

  1. T F F

  2. 略

  3. 参见译文。

Unit 8

Reading Material

  1. T T F

2. A C B B B

3. 参见原文和译文。

Extensive Reading

1. F F T T

2. 参见译文。

Unit 9

Reading Material

1. F T T T T

2. B A A B

3. 略

Extensive Reading

1. T F F F

2. B A A B

3. 参见译文。

Unit 10

Reading Material

1. F T F F

2. C B B A

3. 略

Extensive Reading

1. F F T F

2. 略

3. 参见译文。

Unit 11

Reading Material

1. F T T

2. B C B A

3. 参见原文和译文。

Extensive Reading

1. T F F T

2. 参见译文。

Unit 12

Reading Material

1. F T T F

2. A B C B

3. 略

Extensive Reading

1. F F T T F

2. B A B A B

3. 略

Unit 13

Reading Material

1. F T T T

2. C B C A

3. 略

Extensive Reading

1. F T F F T

2. A B A C

3. 参见译文。

Unit 14

Reading Material

1. T F T F

2. 略

3. 参见译文。

Extensive Reading

1. T F F

2. 略

Appendix C 参考译文

单元 1 人 工 智 能

阅读材料

在计算机科学领域，许多研究都致力于两个方面：一是怎样制造智能计算机；二是怎样制造高速计算机。硬件成本的降低，超大规模集成电路（VLSI）技术的巨大进步以及人工智能（AI）所取得的成果，使得设计面向人工智能应用的计算机结构极为可行，这使智能计算机制造成了近年来最"热门"的方向。

AI 提供了一个全新的方法论，即用计算的概念和方法对人工智能进行研究，因此，它从根本上提供了一个全新的、不同的理论基础。作为一门科学，并且本质上是认知科学的一部分，AI 的目标是了解使智能得以实现的原理。作为一项技术和计算机科学的一部分，AI 的最终目标是设计出完全与人类智慧相媲美的智能计算机系统。尽管科学家们目前尚未实现这个目标，但使计算机更加智能化已取得了很大的进展。计算机已可用来下出极高水平的国际象棋，用来诊断某些类型的疾病，用来发现数学概念，实际上在许多其他领域的表现也都已超出了高水平的人类专业技能。许多 AI 计算机应用系统已成功地投入了实用领域。

AI 是一个正在发展的涵盖许多学科的领域。AI 的分支领域包括：知识表达，学习，定理证明，搜索，问题的求解以及规划，专家系统，自然语言（文本或语音）理解，计算机视觉，机器人和一些其他领域（例如自动编程、AI 教育、游戏，等等）。AI 是使技术适应于人类的关键，它将在下一代自动化系统中发挥关键的作用。

人工智能的一个实际应用是专家系统。专家系统是一个计算机程序，它可以解决人类专家水平的专业问题。

通过将知识库与类似于人类专家的推理能力相结合，专家系统能很好地完成任务。作为人工智能最热门的领域之一，专家系统已被应用于天气预报、医疗诊断、化合物分子结构的确定等领域。

泛读材料

智 能 控 制

一个智能系统应具备在不可预测的环境下恰当工作的能力，在这个环境中，恰当的反应能够增加成功的可能性，该成功是指各个支持系统最终目标的行为子目标的实现。

为了使人造智能系统恰当地工作，它应能模拟生物的功能，最终能够模仿人的智能。一个智能系统能从多个方面来描述。智能程度或水平能从智能的各个方面度量。智能至少要具有感知环境、进而做出决定来控制行动的能力。智能化程度比较高的智能系统具有识别目标和事件、在世界模型中表达知识、思考并规划未来的能力。在智能化程度更高级的形式中，智能具有感知和理解、理智地做出选择、在各种各样的状况下成功运行的能力，以便能在复杂的、不利的环境下生存和发展。通过计算能力的发展和在复杂多变的环境中怎样感知、决策并做出响应的知识的积累，我们可以观察到智能也在成长和发展。

智能系统的上述特征是非常普遍的。据上所述，很多系统都可以被认为是智能的。事实上，根据这种定义，恒温器尽管只是低水平的智能，但是也可以被认为是智能系统。然而，习惯上当一个系统真实具有较高水平的智能时，我们才称它为智能系统。

智能系统具有若干个不同层次的基本属性。人们可以将它们视为智能系统中可用于度量智能程度和水平的特征或维度。下面我们讨论智能控制系统中三种非常基本的特性。

自适应与自学习

智能系统必须具备适应变化的状况的能力。尽管自适应不一定需要自学习能力，但系统要适应不可预测的各种变化，学习能力是必要的。因此学习能力是（高级）智能系统的一种重要特性。

自主性和智能

在设置和实现目标的过程中，自主性是智能控制系统的一个重要特性。当一个系统无需外界干预，能够在不确定的环境中长时间正常运行时，它就被认为是一个高自主性系统。不同系统的自主性程度是不一样的，自适应控制系统被认为比安装了固定控制器的控制系统具有更高的自主能力，这是因为它比固定的反馈控制器能处理更多的不可预测的问题。尽管对低自主性系统来说，无需智能（或"低"智能），但是对高自主性系统来说，智能（或"高"智能程度）是很必要的。

结构和层次

为了应对复杂情况，智能系统必须具备合适的功能结构对控制策略进行有效的分析和评估。这一结构应该是"稀疏的"，并且它应该提供一种机制来构建抽象层次（分辨率、粒度），或者至少提供某种形式的局部排序，以降低复杂程度。具有适应性的层次（或许是大致的、局部的或是组合的层次），可以作为该结构的主要工具来处理复杂性。为了应对多变的环境，自学习能力是必要的，使这种结构能适应重要的、不可预料的变化。

综上所述，智能系统的一个工作特征是：一个智能系统必须对重要的、不可预料的变化具有很高的适应性，因此自学习能力也是必要的。在应对变化时，它必须呈现出高度自主性。它必须能够处理非常复杂的问题，而且这将需要建立某些稀少的、例如层次这样的功能机构。

单元 2　虚 拟 现 实

阅读材料

虚拟现实是指能使一个或多个用户在计算机模拟环境中移动并能够与环境交互的系统。系统中各种各样的仪器能使参与者像操纵真实的物体一样感知和操纵虚拟的物体。这种自然的交互方式能使参与者在虚拟世界中产生沉浸感。这里的虚拟世界是由数学模型和计算机程序创造出来的。

虚拟现实仿真与其他计算机仿真的不同之处在于，它需要特殊的接口设备把虚拟世界的视野、声音及感觉传递给用户。同时，这些设备会记录参与者的语言和动作并发送给计算机仿真程序。

在未来，你的办公室可能不再像一个小隔间，而更像《星际迷航》（美国科幻影视作品）中的全息驾驶舱。计算机科学家们已经在做远程沉浸技术的相关实验，这项技术将使我们看到几百英里外同事的办公室，就像我们与他们正处于同一个物理空间一样。

　　几所大学中正在共同研究这项技术的计算机科学家们已经展示出了一个原型系统，这个系统能使一个在小山上工作的科学家通过两个相对于办公桌安装成直角的屏幕看到离他很远的同事。就好像他从办公室的另一边透过玻璃朝里看一样。与视频会议不同，远程沉浸能提供真实大小的三维图像。

　　在这个原型系统中，每个研究人员身边都围绕着一组可以从不同角度监控他的运动的数码相机。研究人员还戴着头戴式跟踪设备和偏振眼镜，这种眼镜和观看3D电影的眼镜类似。当研究人员头部移动时，他的协同的工作视野也相应地变化。例如，如果他向前倾，他的同事就会显得和他靠得更近，尽管他们仍在几百英里之外。

　　为了使系统运行，高性能计算机必须把每个参与者的数字图像处理成可通过互联网传输的数据，然后将数据重建成图像并投射到屏幕上。到目前为止，因为现在计算机传输数据的速度不够快，运动图像的显示还不连续。

　　科学家们希望远程沉浸技术能开发出大量的其他应用，例如，偏远地区的病人能够"拜访"遥远城市的医疗专家。

泛读材料

未来工厂

　　在现代社会，产品的生命周期变短，为了满足消费者偏好的变化，人们提出了"未来工厂"的概念。产品生命周期缩短，就会产生更具有竞争力的产品，开发出更多新的产品，淘汰掉更多旧的产品，并导致产品订货量的减少。从这个意义上说，大规模生产的时代已经过去，柔性生产的时代正在开始。

　　柔性生产系统的需求确定了未来工厂的规范如下：

1）快速开发新产品。
2）迅速改进具有相似功能的产品。
3）低成本地生产小批量产品。
4）稳定的质量控制。
5）能够生产各种产品。
6）能够生产符合用户特殊修改要求的基础产品。

　　能够满足上述规范的工厂的核心，就是计算机集成制造（CIM）系统。CAD/CAM程序缩短了新产品概念到制造所需要的时间；在主机中安装一个新的程序，FMS就能够生产出新的产品；自动装配线能够适应多种产品的装配问题，这些产品是按照顾客的特定要求进行修改的；自动化检测维持高质量的生产。在车间里只需要少量的工人就能达到所有上述要求。只有物料搬运系统、自动控制装置和工业机器人在人工的定点远程监控下，在车间里进行操作。因此，未来工厂将不设置更衣室、浴室和自助食堂设施。此外，自动化系统、CNC机床和机器人这些CIM系统的基本单元不需要照明和供暖就可以工作。因此，未来工厂将是黑暗和阴凉的。原材料从一端进入，成品将从另一端出来。

　　人们对未来工厂的渴望受到了工业化国家竞争性经济的驱动和现有计算机技术的支持。本文引入的计算机控制概念可使制造系统更加灵活，并能在较短的时间里适应新产品的生产过程。这是朝着实现完整的CIM系统迈出的坚实的一步。看来，美国、日本、欧洲在开发

CIM 系统上所做的努力将使未来工厂不仅仅是一个幻想或梦想，它将在不久的将来变成现实。

单元 3　物　联　网

阅读材料

物联网的基本思想是，这个世界上几乎每一个物体都可以成为连接到互联网的计算机。更准确地说，事物可具有微型计算机的特点，但不会变成计算机。当它们具有这样的特点时，通常被赋予智能化，因为它们比不具备该特点的事物更智能。

环视你周围的电子设备、机器和工具，估算其数量。例如灯泡、汽车、电视、数码相机、冰箱、音响、床，它们都将在未来 10 年内连接到互联网。

这就是"物联网"的潜力：数十亿设备及其组件通过互联网互相连接。物联网将通过"智能"连接从根本上改变我们的世界，节省时间和资源，并提供创新和经济增长的机会。

物联网的基本构建模块是机器对机器（M2M）通信，设备的通信无需人的干预。物联网的一些关键技术为：

1）RFID（无线射频识别）技术。传感器等装置可以随时随地获取信息。

2）互联网的融合能够准确、迅速地传输信息。

3）IPv6 地址。在物联网中，每个物体都应该有一个唯一的地址。这使得物联网成为可能。

我们应该用发展的眼光来看物联网。物联网是一个新兴行业。它将为我们的日常生活带来许多便利。物联网的发展不仅是行业的发展，而且是人民生活的改善。

现在我们正处于物联网技术发展的形成阶段。随着物联网应用的扩展，它将变得成熟并驱动产业链的发展。在未来 3～5 年内，将会建立成熟的应用标准以及行业技术标准。随着产业规模的扩大和传感器技术的有效应用，标准体系将形成初步框架。在未来 5～10 年内，物联网的商业模式将更加活跃。市场将日趋成熟，行业标准将迅速传播并得到广泛的认可。

毫无疑问，中国政府已认识到物联网的重要性。但现在我们正处于发展的形成阶段。在某些领域，政府应该支持和应用这项技术，如用于政策发布、安全管理、能源节约和环境保护。

物联网将通过形成了一个巨大网络的互联网结合各种信息传感设备。其目的是将所有的产品与网络连接在一起，便于识别和管理。物联网不是技术的幻想，而是又一次技术革命。业内专家认为，物联网一方面可以提高经济效益，大大节约成本；另一方面，可以为全球经济复苏提供技术动力。所以我们有理由相信，由于物联网，我们的未来将发生变化。

泛读材料

数据仓库和数据挖掘

数据仓库是一种数据库，它将从各种生产和运营系统中提取的数据合并到一个大型数据库中，可用于管理状况的报告和分析。来源于机构核心事务处理系统的数据被重新组织并与其他信息合并，包括历史数据，这样，这些数据可以用于管理决策和分析。

对于一个不断进取的、具有竞争力的、专注的组织来说，数据仓库不是一个附属品，而是必需品。它是建立决策支持和执行信息系统工具的合适基础，这些工具通常用于衡量和评价组织目标的实现程度。

在大多数情况下，数据仓库中的数据可用来进行报告，不可进行更新，所以公司的基础运营系统的性能不会受到影响。数据仓库这种关注问题解决的特性，使众多公司由于使用了数据仓库而获益。

数据仓库通常具有重构数据的能力。关系数据库的数据视图可使用户从两个维度观察数据。多维数据视图允许用户以多于两个维度的方式观察数据。

数据挖掘是关于使用自动或半自动方法分析数据和寻找隐含模式的技术。数据挖掘对企业有重要的商业价值，比如增加竞争力、客户细分、（客户）流失分析、交叉销售、销售预测、欺诈检测、风险管理等等。

在过去的十年中，大量的数据被积累并存储在数据库中。大部分数据来自商业软件，如财务应用软件、企业资源计划（ERP）系统、客户关系管理（CRM）软件和网络日志。这些数据的收集，使得各个组织的数据丰富而知识匮乏了。数据收集已经受到限制了。数据挖掘的主要目的是从现有数据中提取模式，增加数据的自身价值，并将这些数据转变为知识。

数据挖掘为数据集提供算法，如决策树、聚类算法、关联算法、时序算法等，并分析它们的内容。这种分析产生了模式，可以探索有价值的信息。取决于基本算法，这些模式可以呈现为树形、规则、聚类或仅仅是一组数学公式。在这些模式中发现的信息可用于报告，指导市场营销策略的制定，而最重要的是用于预测。

单元 4　3D 打印

阅读材料

"3D 打印"一词于 1995 年诞生在麻省理工学院，当时的毕业生 Jim Bredt 和 Tim Anderson 改造了一台喷墨打印机，将一种粘合剂挤压到粉末基体上，而不是把墨水挤压到纸张上。之后获得的专利促进了现代 3D 打印企业 Z 公司（Bredt 和 Anderson 合创）和 ExOne 的创建。

3D 打印机工作原理如下。首先，在计算机上调出一个设计图（数字模型），并对其形状和颜色做必要修正。然后你只需按下"打印"键，摆在附近的机器就会开始工作，发出轻微的响声，通过喷嘴注入并沉积材料，或通过胶合剂或者激光选择性地固化薄塑料层或金属粉末层，逐步建立物体。产品就是这样通过逐步添加材料制造出来的，一次铺一层，于是这项技术也叫作"增材制造"。通过这种技术，汽车配件、灯罩，甚至小提琴都可以被打印出来。该技术的优点在于它的运作不一定要在工厂完成，小件物品可以由类似桌面打印机的机器制作，这种机器可以直接摆放在办公室、商店甚至家中的角落；而像自行车车架、汽车面板、飞机零件等稍大的物品则需要更大的打印机和空间。

对比传统的生产方式，这种增材制造具有一些显著优势：通过淘汰生产线而降低生产成本；极大地减少了材料的浪费，用料只占原来的十分之一；它还能够设计制造传统技术难以制造的特殊形状零件，比如飞机机翼、热交换器等；它使单件产品的制造更为便宜、快捷。

目前，3D 打印工序的实现还仅限于特定材料（塑料、树脂和金属），精度大约 0.1 毫米。同 20 世纪 70 年代末期的计算机技术一样，现在仅有少数学术和工业领域的爱好者与工作者在使用 3D 打印。不过正如之前的计算机技术一样，随着技术的进步和成本的降低，3D

打印技术正迅速传播。现在，一台基础3D打印机，也被称为制造者或者"fabber"（一款免费、开源的快速成型机的名字），其价格低于1985年一台激光打印机的价格。

该技术将不仅对资本和就业产生影响，也将触动知识产权（IP）的规则。当任何产品都可以用电子文件描述时，它们将更容易被复制和传播（自然也就更容易遭遇盗版），看看音乐产业所受的冲击就知道了。当一个新玩具的设计，或者一款新鞋的设计泄露到互联网上，设计者丢失知识产权的概率就更大。

就像人们难以预测1750年的蒸汽机、1450年的印刷机，或者1950年的晶体管所带来的影响一样，而今我们也无法预测3D打印将来如何改变这个世界。但是这项技术已经来临，并将改变它所触及的每一个领域。公司、监管部门和企业家们是时候对此做出思考了。至少现在有一件事情是明确的：尽管3D打印技术可能在短期内造成"几家欢喜几家愁"的局面，但长远来看，它必将开拓新的行业领域。

泛读材料

先进制造技术（AMT）

技术创新带动着人类社会的进步，近几年世界的变化更是日新月异。瞬息万变的市场要求保证产品质优价廉的同时大幅缩短产品生命周期。现在，用户都希望有更多种类的产品可供选择。这种现象也激励着制造企业在各个工艺流程中追求计算机自动化，多样化小批量生产也正逐渐替代大规模生产。制造企业已经认识到面对多元化市场带来的挑战时柔性制造系统的重要性。对于应对各种产品更短的交货周期、更高的生产效率和质量要求的各种挑战，柔性制造系统的概念起着至关重要的作用。柔性制造是传统生产方法向更加集成化、自动化生产方法过渡的基本概念。这些方法强调的是实施自动化生产项目的企业应该将长期战略规划中的柔性要求优先落实。

关于AMT的定义有很多种。例如，Baldwin（1995）把AMT定义为一组基于硬件和软件的集成技术，如果恰当地实施、监控和评估，将提高公司制造产品或提供服务的效率和效益。广义上说，AMT是一个完整的社会技术系统，由所采用的方法定义技术的融合水平。它囊括了计算机辅助制造（CAM）、柔性制造系统（FMS）、制造资源规划（MRP）、自动化物料处理系统、机器人技术、计算机数控（CNC）机床、计算机集成制造（CIM）系统、最佳生产技术（OPT）、准时制生产方式（JIT）。尽管AMT非常注重技术创新的运用，但由于AMT系统需要不间断的检查、调整，管理也起着举足轻重的作用。

先进制造技术（AMT）的内在优势会为企业，特别是地区性小企业创造新的发展机会。AMT可以快速、准确地调整生产规格，这意味着企业可以为他们的产品提供定制服务，并依靠低产量、低成本盈利。传统意义上讲，技术只是实施业务战略的一种工具，而AMT有可能直接影响企业的战略选择。到目前为止，AMT资料表明采用AMT的企业会拥有不断追求战略创新的能力。

单元5　工业机器人

阅读材料

机器人是一种可重复编程的多功能机械手，其设计用途是通过各种程序化动作来移动物

Appendixes

料、工件、工具及特殊装置，从而完成各种不同的任务。工业机器人是一种提高制造生产力的工具。它可以承担那些对人类可能有危险的工作，最早的工业机器人就曾经用来在核电厂中更换核燃料棒。工业机器人也能在装配线上工作，如安装印刷电路板上的电子元件。这样，就可以将工人从这种单调工作的常规作业中解脱出来。机器人还能通过编程拆除炸弹，为伤残人士服务，为我们的社会做各种各样的工作。

与常规观点相反，机器人学并非最近才发展起来的。事实上，早在20世纪60年代初期，第一批机器人便在美国被创造出来。Unimation（第一家机器人制造公司）1961年就生产出了一款机器人手臂，其控制装置的时序是由操作者预设的。然而，鉴于这项工作当时尚属试验，为了避免公众的抵制情绪，该项目保持了低调处理。1974年，Cincinnati Millicron机器人成为首台小型计算机控制机器人。然而，就在同一年，瑞典公司ASEA推出了IRB6机器人；这一系列的机器人一直在全球畅销，至今还在生产，相应的重大改进是控制柜电子装置与软件的升级。所以，当人们以为是美国建立了机器人技术时，日本和瑞典等国家在工业中的机器人应用已经达到了很高的水平。当前，对于机器人的研究集中在制造"智能型"机器人，集"视""听""触摸"等功能于一身，进而能做出决定。

为了划分某种工业机器人的设计属于何种类型，我们必须能够识别其运动结构。与人类手臂相似，机器人手臂由一系列连杆与关节组成。关节是手臂中连接两个连杆的部件，它允许各连杆之间存在相对运动。为了确定运动结构，我们需要知道所使用的关节类型与手臂的"自由度"。

每个机器人都有一个基座，该基座一般固定在地面上。但是，基座也可以固定在墙壁或者天花板上，也可以安装到龙门架上。所有这些都使机器人具有节省地面空间的优点，并改善了机器人的工作范围，从而增强了终端执行器（一个描述机器人手臂终端工具的通用术语）的操作能力。机器人手臂的第一个连杆与基座相连，最后一个连杆与终端执行器相连。一般而言，机器人手臂上的关节越多，其动作就越灵活。工业机器人的典型结构由机械手、终端执行器、驱动系统和控制系统四个主要部分构成。

泛读材料

自动生产线

使用自动生产线可以实现专用、多功能机床最大程度的自动化。自动生产线实质上是各个独立工作站的组合体，这些工作站按需求顺序排列，由工件传送装置连接，并且通过连锁控制集成。工件在工位间被自动传送，每个工位都装配有用于加工、测量、工件再定位、组装、清洗及其他操作的卧式、立式及倾斜式的设备。自动生产线的两大主要类别是旋转式和直列式。

自动生产线的一个显著优点是它们允许同时完成大量的操作。相对来说，对工件可加工表面或平面的数量没有限制，这是因为自动生产线几乎可以在任何节点接入装置，执行工件翻转、旋转或定向工序，以便完成加工操作。工件再定位也使倾斜主轴箱的使用需求降至最低，并使操作在最佳时间完成。从原始铸件或锻件一次性完成成品件的加工通常是可行的。

所有类型的机加工操作，如钻削、攻丝、铰孔、镗削和铣削，在自动生产线上被经济地组合在了一起。诸如车削和表面加工的车床式操作也可在直列式自动生产线上完成，工件在特定的机加工工位上旋转。车削操作在机床上完成，车刀架通过安装在隧道式桥接单元上的滑轨进给。在每个车削工位上,工件定位在中心位置,并由卡盘带动旋转。CNC 车削工位可用于直列式自动生产线。CNC 装置使得机器工作周期便于调整,以适应工件设计的改变,也可用于自动刀具的调整。

当工件在生产线上移动时将零件组装到工件上,经常可以获得生产线最大的生产经济效益。在自动生产线加工过程中,能够对轴衬、密封垫、威尔士衬套和热管等零件进行组装、机加工或测试。完成零部件装配后也可自动进行螺母安装。

如果能使用合适的机加工装置且随后进行良好的操作,在自动生产线上进行深钻或铰孔是一项理想的应用。球面座和其他表面的仿形镗削和车削可采用示踪控制的单点插入加工法完成,从而无需昂贵的专用成形刀具。对铰孔或镗孔的过程测量以及刀具的自动调整是在自动生产线上进行的,以保持精确的公差。

长期以来,自动生产线一直用于汽车工业以提高生产效率,手工零件加工量极少。除了减少劳动力需求外,自动生产线可始终保证以更低的成本生产符合标准的优质零件。它们不再局限于粗加工,现在通常不再需要诸如研磨和珩磨这样的后续工序。

单元 6　计算机集成制造系统

阅读材料

CIMS 是计算机集成制造系统 "Computer Integrated Manufacturing System"的缩写。它是随着 CAD 和 CAM(计算机辅助设计与制造)的发展而产生的。CIMS 描述了一种制造、管理和企业运营的新方法。虽然 CIM 系统可能包含许多先进的制造技术,例如机器人技术、计算机数字控制(CNC)、计算机辅助设计(CAD)、计算机辅助制造(CAM)、计算机辅助工程(CAE)和准时制生产等,但它超越了这些技术。

作为一个全自动系统和计算机软件,CIMS 制造涵盖了 CIMS 的三个主要领域,包括工厂自动化、生产和工艺设计、制造规划和控制。

CIMS 包括完成全部 CIMS 工艺所需的软件和自动化系统。具体工艺包括产品设计、系统编程、生产成本估算、产品的实际制造、订单输入、库存跟踪以及实际制造成本分析环节。

CIMS 是一种真正的柔性制造系统,不需要对系统做结构上的改变,就可以小批量地制造各种各样的零件和部件。之所以可行是因为通过使用软件,编程人员可以很容易地对物料传输路径、机器人程序、CNC 机床程序、自动存储和检索系统进行编程。在程序的控制下,自动存储和检索系统可提供许多不同的零件。最后,柔性传送装置利用输送带上的托盘,并通过使用可互换的工具处理多种不同的材料。

泛读材料

计算机数控机床

从最简单的机床到最复杂的机床都会用到 NC(数控技术)。数控机床是在普通机床的基础上发展起来的,各种类型的数控机床都来源于同类型的普通机床。经过 40 多年的发展,数控机床的规格、型号繁多,而且结构和功能也各具特色。为了了解数控机床,我们简单地介绍几种主要的计算机数控机床。

Appendixes

1. 数控车床

图 6-3 所示是一台数控车床的外观图。数控车床是生产效率最高的机床之一。数控车床主体包括主轴、溜板、刀架等。数控系统包括 CRT 显示器、控制面板以及强电控制系统。

数控车床一般具有两轴联动功能，Z 轴是与主轴平行的运动轴，X 轴是在水平面内与主轴相垂直的运动轴。Z 轴控制机床刀架靠近或离开主轴箱的运动，X 轴控制刀具的横向运动。在最新的车铣加工中心中增加 C 轴，可实现工件的分度要求，在刀架中安装铣刀，可对工件进行铣削加工。

2. 数控铣床

数控铣床是工业生产中加工方式最多的机床之一，像铣削、成形加工、齿轮加工、钻削、镗削以及铰孔只是其众多加工方式中的一小部分。数控铣床适用于加工三维复杂曲面，在汽车、航空航天、模具等行业中被广泛采用。世界上第一台数控机床就是数控铣床，但随着时代的发展，数控铣床的趋势是向加工中心转变。由于具有较低的价格、方便灵活的操作、较短的工作准备时间等优势，目前数控铣床仍被广泛使用，它可分为数控立式铣床、数控卧式铣床和数控仿形铣床等。

3. 加工中心

加工中心是数控机床发展到一定阶段的产物，至今人们对加工中心还没有明确的定义，装配有自动换刀装置的数控镗铣床，通常被认为是加工中心。实际上，加工中心可以概括为："具有自动换刀装置，并能进行多种工序加工的数控机床"。加工中心可以进行铣、镗、钻、铰、攻丝等多种工序的加工，这就减少了对单个普通机床的需求，因而减少了资金和劳动力需求。

加工中心可以分为立式加工中心和卧式加工中心，立式加工中心的主轴是垂直方向的，卧式加工中心的主轴是水平方向的。立式加工中心一直被广泛接受和使用，它主要用于加工平板零件和需要在一个零件表面进行三坐标加工的场合，比如模具和型腔类工件。卧式加工中心也被广泛接受和使用，尤其适用于加工大型的、箱体的、重型的零件，这是因为当用于单元制造或柔性制造系统时，卧式加工中心可以适应简单易行的托盘往复传送。

工件可以通过夹具安放在回转工作台上或交换托盘上，通过工作台的旋转可加工多面体，托盘交换可更换加工的工件，从而提高加工效率。

4. 数控钻床

图 6-6 所示是一台数控钻床。数控钻床可分为数控立式钻床和数控卧式钻床。数控钻床可进行钻孔和攻丝作业，同时也可以完成简单的铣削作业，它的刀库可以存放多种刀具。

5. 数控磨床

数控磨床主要用于高硬度、高精度表面加工。数控磨床可分为数控平面磨床、数控内圆磨床和数控仿形磨床。随着自动砂轮补偿技术、自动砂轮修正技术和磨削固定循环技术的发展，数控磨床的功能越来越强。

6. 数控电火花成型机床

每一位机械师都知道，在标准机床上，电能通过电动机转换成动能。现在，人们发现电能可以直接用于切除金属。数控电火花成型机床是一种特种加工机床，它利用两个不同极性的电极在绝缘体中产生放电来切削材料，进而完成加工。数控电火花成型机床在加工复杂模具和难加工材料方面有独特的优势。

7. 数控线切割机床

数控线切割机床的工作原理与数控电火花成型机床一样，其电极是电极丝，加工液一般采用去离子水。

单元 7 机电一体化技术

阅读材料

1969 年，安川电气公司的一位日本工程师创造了"mechatronics"（机电一体化）一词，以反映机械与电气工程学科的融合。直到 20 世纪 80 年代初，机电一体化一直指的是电气化的机械装置。在 20 世纪 80 年代中期，机电一体化产生了工程学的概念，指机械学和电子学的分界学科。现在，机电一体化涵盖了大量的技术概念，这些技术都是各自领域内的知名技术。每项技术仍然以机械和电子技术的融合作为基础，但现在很多技术已包括了更多的新兴科技，特别是软件和信息技术。例如，许多由机电系统产生的早期机器人已经成为机电一体化技术的核心。

机电一体化是机械学、电子学、信息技术等多学科融合的学科，其目标是使工程师完成产品开发，这就是为什么机电一体化技术目前在工业中很受欢迎。随着科学技术的不断发展，机电一体化发展至今也已成为一门有着自身体系的新型学科，但不仅仅是被赋予新的内容这样简单。其基本特征可概括为：机电一体化是从系统的角度出发，综合运用机械技术、微电子技术、自动控制技术、计算机技术、信息技术、传感测控技术、电力电子技术、接口技术、信息转换技术以及软件编程技术。根据系统功能目标和优化组织目标，各功能单元的合理配置与布局，在多功能、高质量、高可靠性、低能耗的意义上实现特定功能价值，并使整个系统最优化。由此而产生的功能系统，则成为一个机电一体化系统或机电一体化产品。

在工厂自动化方面，机电一体化技术在制造中产生了深远的影响，而且将来还可能变得更为重要。工厂自动化的主要组成部分包括 CNC 机床、机器人、自动化系统以及 CIMS。这些先进制造方案基本上是由机电一体化系统组成的。未来的用户需求可能是体积小、品种多，灵活性更高，加工周期短，制造和装配自动化。显然，未来产品的设计和制造将会涉及精密机械与电子系统的结合，机电一体化将成为所有产品和生产技术相关活动的核心。

泛读材料

可编程逻辑控制器（PLC）概述

可编程逻辑控制器（PLC），也被称作可编程控制器，是计算机家族中的一员。它们被应用于商业和工业领域。PLC 通过监控输入、根据程序做出决策、控制输出来自动控制过程或机器。本文旨在向您提供 PLC 功能和配置的基本信息。

PLC 通常包括输入模块或输入点、一个中央处理单元（CPU）以及输出模块或输出点。输入单元可以接收来自不同现场设备（传感器）的各种数字或模拟信号，然后将它们转换为 CPU 可以使用的逻辑信号。CPU 根据存储器中的程序指令做出判断，然后执行控制指令。输出模块将来自 CPU 的控制指令转换为可用于控制各种现场设备（执行机构）的数字或模拟信号。编程设备用于输入期望的指令，由这些指令来决定 PLC 对于特定输入的反馈。操作界面能够显示过程信息和输入新的控制参数。举一个简单的例子，连接到 PLC 输入端的按钮（传感器），通过电动机的起动器（执行机构）可以起动和停止与 PLC 相连的电动机。

使用 PLC 之前，许多控制任务是通过接触器和继电器控制来完成的。这就是经常说的硬接线控制。硬接线控制必须先设计电路图，确定和安装电气元件，创建接线表，然后电工再连接执行特定任务需要的元件。如果期间出现错误，就必须重新连线来纠正。改变功能或扩充系统需要大量的元件更改，并且线路要重接。

同样的以及更加复杂的任务，可以由 PLC 来完成。设备和继电器触点之间的连线在 PLC 程序中完成。虽然依然需要连接现场设备，但硬接线密度更低。修改应用和纠正错误要容易操作。在 PLC 中建立和修改程序比连线和重新连接电路要容易。

下面列出的只是 PLC 的一部分优势：
1) 比硬接线方案的物理尺寸小。
2) 修改更简单、更快速。
3) PLC 综合了诊断和超控功能。
4) 应用可以及时记录。
5) 应用能够更快地、以更低的成本复制。

单元 8　多传感器数据融合

阅读材料

多传感器数据融合提供了一种提高单个传感器性能的途径。一般来说，从单个传感器很难得到高质量的测量数据。如果利用多个传感器执行相同的测量，再将这些传感器的测量数据以某种方式组合起来，那么结果将很可能超过每个单独测量结果的精度。

总之，顾名思义，MDF 是一项将多个传感器的数据通过一个中央数据处理器进行组合，以提供全面而准确的信息的技术。MDF 的应用领域广泛，军用方面如目标自动检测与跟踪、战场侦察等；民用方面如环境监测、复杂机械监控、医疗诊断、智能建筑、食品质量检验以及精细农业。

在测量与仪器的数据分析和处理中，模式识别技术是必要的。模式识别可用于开发数据融合算法。人工神经网络，基于对人脑机制的研究而开发，是胜过其他常规统计模式识别方法的首选。Linn 和 Hall 调查研究了 50 多个数据融合系统。其中仅 3 个系统使用了神经网络方法。如此低的数据可能表明神经网络在数据融合领域的重要性被低估了。

人工神经网络已被广泛用于解决复杂问题，例如模式识别、快速信息处理和自适应。人工神经网络是基于对人脑机理和结构的研究而构造的。神经网络的结构和实现就是人脑结构和活动的模拟简化版。生物神经组织固有的强大处理能力启发了人们对其结构本身的研究，并将其作为设计和组织人工计算结构的模型。集成了神经网络模式识别的 MDF 体系是一种前景广阔，可实现测量与仪器中高质量数据分析和处理的结构。

泛读材料

质 量 控 制

根据美国质量管理协会（ASQC）的定义，质量是产品或服务能够满足规定需求而具有的特性和特征的总和。这个定义表明，必须首先确定顾客的需求，因为满足这些需求是实现质量的"底线"。然后应该把顾客的需求转化为产品的特征和特性，据此开展设计工作和制订产品的技术要求。

质量正在迅速成为顾客挑选商品和服务的重要影响因素。质量控制（QC）功能一直沿用传统的人工检测方法和统计分析程序。人工检测通常是一种耗时的工序，并涉及精度要求高，但又比较单调的工作。它经常要求将零部件从防护设备中运送到一个独立的检测区域检测，这样就会引起延误，常常还会影响生产计划。

使用统计抽样程序的前提是承认存在一些缺陷零件不会被发现的风险。实际上，统计质量控制的目的在于确保在生产检测过程中能够达到预计的或平均水平的零件缺陷率。传统统计质量控制程序的本质是必须接受某些合格率小于100%的产品。

传统质量控制监测工序从另一方面讲，会减损效益，这通常在产品制造完毕后才体现出来。在零件制造完成之后，会对零件进行检测并进行质量评定。当某些零件检测不合格时，往往会以高于其原始制造成本的成本报废或返工。

质量控制过程包含从供应商，到生产过程，再到客户的所有活动：检查原材料以确保它们符合相应的规范；对半成品质量进行分析以确保生产过程的正常运行；对成品及服务进行研究以确定它们能否满足客户期望。伴随着自动化，检验和测试的成本降低、时间变短，因此公司可以增加样本量和提高抽检频率，从而使控制图和验收计划更为精确。

单元 9　自动识别技术

阅读材料

自动识别技术是指在没有人为干预的情况下自动收集目标数据、并将数据存入数据库的任何相关技术。自动识别技术无处不在，安静而高效地执行数以千计的平凡工作。最适用的一项重要工作是通过自动识别技术回答一些信息交流方面的问题，如"它是什么？""它在哪里？""它怎么样？"

自动识别技术主要用于识别和跟踪箱体、人、动物等，但凡能叫出名字的物体都可以识别和跟踪。与人相比，自动识别技术能够更快、更准确，并以更低的总成本进行识别和跟踪。自动识别技术有很多种类，如磁墨水字符识别（MICR）、磁条、语音识别、生物识别和条形码等，而 RFID 仅是其中之一。

既然有这么多种自动识别技术，为什么像 RFID 这样的技术还会突然变得如此受欢迎呢？所有都归结于一点，即无线电波。RFID 技术应用电磁波（无线电波）（电磁波谱的一部分）来识别单项商品、地点、动物或人类。

RFID 的用途很多，最常见的用途是使用识别码（类似名字）唯一地标识需要识别的商品、地点、动物和人，该识别码存储在与天线相连接的集成电路（IC）中。集成电路和天线统称为 RFID 发射器或标签。标签要与需识别的商品、地点、动物和人相连。与标签进行通信并从标签中读取识别码的设备为问询器或读卡器，读卡器将读取的识别码输入到信息系统中，信息系统将识别码储存在数据库中，或从数据库中找出识别码，然后将存储的关于商品、地点、动物或人的信息返回。各种自动识别技术的主要区别为存储和读取识别码的方式不同。

泛读材料

生物信息学

在此，我们将探究由试图确定计算机技术如何推进生物学发展以及生物学又是如何使计算机科学更有活力而引出的许多问题。要花多大努力才能使计算机科学家们改变方向去做一些生物信息学的工作呢？生物信息学的哪些主题与计算机科学相接近呢？计算机科学研究能否以及如何在生物学中得到灵感仍然是一个长期的命题。因而，就必须探索一些与此相关的话题，包括：生物学在整个科学领域的迅速传播；计算机科学和分子生物学的文化差异；分子生物学的当前目标；生物信息学用到的网络数据；生物学家感兴趣的计算机科学领域。

自从1953年James Watson和Francis Crick因确定DNA结构而取得了里程碑式的成就以来，生物学，尤其是分子生物学，取得了突飞猛进的发展。人类基因组测序就是诸多成就中的一个代表。随后，许多其他生物体的基因组测序也被研究出来。没有计算机的帮助，大部分成就是不可能实现的，这也引出了一些问题，包括：计算机在生物学中的角色是什么？是像把人类送上月球一样，是一个工具，还是仅是众多工具之一呢？分子生物学本质上是一门分离的学科，这是它与计算机技术的共性。

当一个领域与另一个领域融合，带有连字符的术语所表达的含义通常是含糊的。比如说，生物物理学是属于物理学知识主体还是生物学知识主体呢？这是一个独立的新学科吗？大概随着时间的推移和成就的取得，新学科的解释会站得住脚。数学生物学需要的数学知识比生物知识要多。同样地，计算机科学家正在开发的计算生物学是用来满足生物学家的需求，但基本上需要大量的计算机科学理论知识。

生物学家对于用来解决他们日常问题的计算机科学理论不会特别感兴趣。术语"生物信息学"对生物学家更有吸引力，因为他们对于"计算生物学"中的"计算"一词不那么感兴趣。生物信息学是一门发展中的跨学科科学。其他学科的参与（如计算机科学）使得该学科前景广阔。不管结果如何，计算机科学家肯定会和生物学家成为积极的、坚定的合作者并从中受益。

单元10　太空科学探索

阅读材料

最近几年，在运用大功率火箭将仪器送入远离地球的高空以发射人造卫星和深空探测器方面，人们做出了越来越多的努力。我们已经指出了人们在地球表面已经做了多少，还能做多少。那么，为什么所有这些都集中于火箭的运用呢？

主要原因之一是我们的大气层，尽管总体上来说大气层的存在是有利于生命生存的，但它阻碍我们在非常有限的光谱范围之外的波域观察太空，实际上，可观察波谱几乎完全被限制在可见光和相对有限的无线电波范围之内。我们必须从大气层外进行观察，来研究紫外线、X射线、红外线和所有无法穿越大气层的无线电波。借助于离地球200多英里高空运行的人造卫星上的仪器，可以进行这种观察。我们不知道这些仪器会记录些什么，如果知道的话，就不值得找这么多麻烦了，但是天文学领域有待于以后几代人进行研究。

在地球外部大气层研究、气象学、地球与行星间的空间研究等领域中，火箭运载工具的

科 技 英 语

发展开辟了很多科学研究的新的可能性，光线研究只是应用之一。在称为空间探索的工作中，有四种主要运载工具。首先是垂直探测火箭。其次，我们拥有以椭圆形轨道绕地球运行的人造卫星。如果卫星轨道特别长，超过地球半径（4000英里）数倍时，我们就采用深空探测器。探测器可以经过特定导向从月球附近掠过，也可撞击月球，或成为月球的卫星。这就是月球探测器，近年来我们已经有了许多这类探测器的应用实例。

从科学家的观点来看，所有这些运载工具都很有价值。任何一次特定发射的价值都代表着相关试验的成功，而不只是离开地球的距离长短。科学家也不关心这些运载工具是否载人，因为仪器可以自动操作并将其记录的数据转换成无线电信号，从数百万英里以外送回到地球。

泛读材料

神舟飞船

神舟飞船由轨道舱、返回舱和推进舱三个主要的舱体结构组成，另外还有一个附加段。

轨道舱是飞船在地球轨道上运行时航天员生活和进行各项科学实验的地方。返回舱是航天员在飞船发射和返回地球阶段所待的地方。而推进舱则用来为飞船提供动力，控制其位置，改变其轨道并使飞船返回地球。

轨道舱位于飞船的前部，呈圆柱形密闭结构，两边装有太阳能电池，其外部装有配有各种天线的太阳能传感器。轨道舱设有飞船对接操作需要的对接口。没有对接计划时，对接口由附加的材料密封。

整个飞船的独特之处就在于轨道舱具有多种用途：飞行过程中，它是航天员的起居室；还是航天员的实验室；如果需要对接其他飞船，它还是对接的目标舱体；在航天员进行太空行走时，它是一个空气制动器；飞船返航时，它又是一个支撑体。

神舟飞船在结构上与最早的苏联和美国的载人飞船有很大区别。将人类历史上第一位宇航员，苏联宇航员 Yuri Gagarin 送上太空的东方号飞船，以及载着美国第一名宇航员 Alan Shepard 的自由水星7号飞船都只有两个舱体，而神舟飞船却有三个。

在第一次太空飞行中，东方号只绕着地球转了一圈，水星号也只进行了一次近地飞行。相比之下，神舟飞船载着中国航天员在近地轨道上飞行了近一天的时间。神舟飞船主要依靠太阳能电池供电，这比水星号和东方号上的动力系统更为先进。

单元 11 微 处 理 器

阅读材料

微处理器是一种超大规模集成电路。集成电路，也被称为微型芯片或者芯片，是复杂的电子电路，由单个、薄而平整的半导体材料上形成的极其微小的元件组成。除了电阻、二极管、电容和导线等其他元件之外，现代微处理器还集成了多达1千万个晶体管（用作电子放大器、振荡器，或最常见的是用作开关），上述所有元件通通被封装在大约一枚邮票大小的区域内。

一个微处理器由几个不同的部分组成：算术逻辑单元（ALU）实现数字计算和逻辑决策；寄存器是存储临时信息的特殊记忆区，就像便签本的功能一样；控制单元解读程序；总线传输芯片和计算机中的数据信息；本地内存支持片上计算。更复杂的微处理器往往包含其他部件，例如专用存储区，也叫作高速缓冲存储器，可以加速微处理器对外部数据存储设备的访问。现代微处理器是以 64 比特（比特是二进制数，用 1 和 0 表示的信息单元）的总线宽度工作的，这就是说，可同时传输 64 位数据。

制造微处理器采用的技术与制造其他集成电路使用的技术类似，例如制造存储芯片。微处理器通常具有比其他芯片更复杂的结构，而且微处理器的制造需要极其精细的技术。微处理器的经济制造需要批量生产。在硅晶片的表面，上百个模具或电路样板被同时制造出来。微处理器是通过导电材料、绝缘材料和半导体材料的沉积和去除工艺制造的，每次一薄层，经过几百个独立的工艺步骤后，才制成一个包含了微处理器所有互联电路的复杂夹层结构。

泛读材料

数字信号处理

数字信号是一种由 1 和 0 组成的、能用数学方法处理的语言。对应我们在现实世界所讲的模拟信号。模拟信号是我们每天都感受到的现实世界信号，如声音、光、温度和压力。数字信号是模拟信号的数值表示。在数字世界里，对数字信号进行处理可能会更容易、更节约成本。在现实世界中，我们可以通过模数转换器将模拟信号转换为数字信号，然后对数字信号进行处理；如果需要的话，可以用数模转换器将信号转换到模拟世界中去。

信号既可以使用模拟技术（模拟信号处理，ASP）进行处理，也可以使用数字技术（数字信号处理，DSP）进行处理，还可以使用模拟和数字技术（混合信号处理，MSP）进行处理。

数字信号处理器是信号处理的核心。数字信号处理器（DSP）是一种速度极快、功能强大的微处理器。说到 DSP，它与传统计算机数据分析的区别在于，它在实时执行滤波、FFT 分析和数据压缩等复杂信号处理功能时的速度和效率更快、更高。

DSP 之所以与众不同，是因为它能够实时地处理数据。实时处理能力使得数字信号处理器非常适用于那些不容许任何延迟的场合。例如，你是否使用过两人无法同时讲话的手机呢？你不得不等对方说完才能开口。如果你们两人同时讲话，信号就会掉线，你听不到对方说话了。今天的数字手机使用了 DSP，这样人们就可以正常通话。手机中的 DSP 处理声音的速度如此之快，你在说话的同时就可以实时听到。与其他微处理器相比较，使用专用数字信号处理器进行设计有如下几点优势：

1）单周期乘法累加运算。
2）实时性能模拟和仿真。
3）灵活性。
4）可靠性。
5）系统性能提高。
6）系统成本降低。

数字信号处理技术及其应用正以惊人的速度发展。随着大规模集成电路的出现和由此带来的数字元件的成本下降、体积缩小和运算速度提高，数字信号处理技术的应用范围不断扩大。目前，专用数字滤波器的取样率高达兆赫兹。数字信号处理器也成为许多现代雷达和声纳系统不可缺少的组成部分。

单元 12　未来汽车和自动驾驶

阅读材料

　　在未来汽车中，我们将拥有自动驾驶开关。ABS（防抱死制动系统）是汽车上第一个能针对驾驶员命令背道而行的系统。ABS 会违背驾驶员的意愿，减少某个车轮上的制动力。其结果是，电子系统介入判定，使汽车更加安全。安全气囊也由电子系统判定是否膨胀。

　　必须要迅速做出一个简单的决定时，计算机将接管任务。今天，我们的技术水平能获得由计算机解译的三维图像；我们可以用电子仪器测量距离。计算机可对高速公路进行监控。这些基本工具可用于高速公路上的自动驾驶。问题是，计算机驾驶的汽车不允许犯错误，否则制造方将要负责。所以实现全面自动驾驶还将需要一段相当长的时间。

　　（在未来汽车中，）类似 ABS 的安全系统将首先被使用。渐渐地，像巡航控制系统和 GPS（全球定位系统）等舒适性系统也将会应用于汽车，巡航控制系统能够适应外部的因素，如天气、路标，以及我们周围的交通状况等。

　　尽管我们会乐意有一个系统帮助我们避免类似追尾的交通事故，但自动驾驶技术是非常复杂的。

泛读材料

电动汽车和混合动力汽车

　　在汽车史上，人类曾试验过蒸汽机、燃气涡轮机、转子式发动机和太阳能动力的汽车。但所有这些都不能与内燃活塞式发动机的动力和效率相媲美，内燃机由矿物燃料驱动，并与通过一系列轴和齿轮驱动车轮的变速器相连。

　　不幸的是，由于存在燃料短缺、空气污染等令人困扰的问题，必须重新搜寻替代能源，以满足人们的交通需求。虽然工程师们一直在潜心研究新的动力系统，使其效率远远超过我们现在使用的动力系统，但问题是，这些技术大都依赖于汽油之外的燃料。

　　众所周知，内燃机汽车废气中的 CO、HC 及 NO、微粒、臭氧等污染物会形成酸雨、酸雾及光化学烟雾等。

　　电动汽车（EVs）由电动机驱动，电动机的动力来自可充电电池组。电动汽车不像内燃机汽车那样，工作时不产生排气污染，这对保护环境和空气的洁净是十分有益的，所以电动汽车有"零污染"的美称。

　　电动汽车由电力驱动及控制系统、机械传动系统和执行装置等组成。电力驱动及控制系统是电动汽车的核心，也是区别于内燃机汽车的最大的不同点。电力驱动及控制系统由驱动电动机、电源和电动机的调速控制装置等组成。而电动汽车的其他基础装置与内燃机汽车相同。

　　混合动力电动汽车（HEVs）为消费者、车队和国家带来了好处。这些先进的车辆能减少燃油的使用和成本，同时保持车辆性能，保护公众健康和环境，并增强能源安全。

　　混合动力电动汽车的排放量取决于车辆及其配置。一般而言，混合动力电动汽车比同类传统汽车的排放量低，这是因为电动机可一定程度上减少内燃机的使用。此外，混合动力电动汽车还可以在纯电动模式下运行，在这种模式下，车辆零排放运行，在人口密度大和不允许有排放的地方，这是最佳选择。

混合动力电动汽车与传统车辆一样，需要常规保养。建议这些车辆由合格的经销商进行保养，以确保维修保养有保障。

混合动力电动汽车是非常安全的车辆。它们会接受与传统车辆一样严格的测试且必须符合同样的安全标准，包括碰撞试验和安全气囊。

单元 13　计算机网络

阅读材料

计算机网络，通常简称为网络，是由通信信道连接的计算机和设备的集合，它可促进用户之间的通信和资源共享。计算机网络由计算机、打印机和其他连接在一起的设备组成，设备之间能够互相通信。图 13-1 给出了一个校园局域网的例子，它是将计算机、因特网以及各种服务器连接在一起。

计算机网络具有多种用途：

便利通信。利用网络，通过电子邮件、即时消息、聊天室、电话、视频电话和视频会议等方式，人们可以高效、方便地沟通。

共享硬件。在网络环境中，网络上的每台计算机都可以访问和使用网络上的硬件。假设网络上的几台个人计算机每台都需要使用激光打印机，那么只要把这几台个人计算机和激光打印机都连接到网络上，用户就可以根据需要访问网络上的激光打印机了。

共享文件、数据和信息。在网络环境中，任何授权用户都可以访问网络上其他计算机中存储的数据和信息。能够提供共享存储设备中数据和信息的访问是许多网络的重要特征。

共享软件。与网络连接的用户可以访问网络上的应用程序。

网络中的计算机连接可以是点对点、媒介共享、无线式，也可以是这些形式的组合。最简单的形式是点对点连接，两台计算机直接由一条电缆连接，可以通过连接的电缆实现信息的写入和读出，即信息的传递。更为普遍的是，网络计算机之间的连接要求网络中所有计算机能共享同一通信媒介。此媒介可以是一条电缆，也可以是无线电、微波或卫星传输信道。

泛读材料

信　息　安　全

信息安全是指保护信息和信息系统，阻止未经授权的访问、使用、披露、中断、修改或破坏。对于因特网上的信息，至关重要的三个基本安全概念是要保证其保密性、完整性和可用性。对于信息的使用者，要进行认证、授权、确认。

信息被未经授权者读取或复制即构成失密。对于某些类型的信息，保密是非常重要的。具体示例包括研究数据、医疗和保险记录、新产品说明书、企业的投资策略等。在某些地方，保护个人隐私可能属于法律义务。对于银行和贷款公司、追债公司、给客户提供信贷或发行信用卡的商行、医院、医生办公室、医疗检测实验室、提供心理辅导或药物治疗的个人或机构、收税机构等，保护个人隐私尤其重要。

不安全网络上的信息可能会遭到破坏。当信息以意想不到的方式被修改时，其结果就是丧失了它的完整性。这意味着无论是人为错误还是蓄意篡改，信息都遭到未经授权的更改。

科技英语

保持信息完整性对关键安全事务和财务会计是尤其重要的。

在不安全的网络环境下，未经授权访问信息是非常容易的，而且很难抓住入侵者。即使用户在自己的电脑中没有储存他们认为很重要的信息，该计算机也会成为"薄弱环节"，未经授权的入侵者能够访问其系统及信息。

看似无害的信息能够暴露一个计算机系统，使其受到安全的威胁。入侵者能发现的有用信息有系统正在使用的硬件和软件、系统配置、网络连接类型、电话号码、重要文件和程序访问，从而影响系统的安全性。如密码、访问控制文件和密钥、人员信息和加密算法等即为重要信息。

单元 14　电子商务

阅读材料

电子商务又称电子贸易，是指通过计算机网络（如互联网）以及电子化的交易方式进行采购、销售、互换商品及服务的过程。与人们的普遍认识相反，电子商务不只是通过网络开展交易。事实上，在网络尚未出现的20世纪70年代，电子商务就已经存在，并且运用EDI（电子数据交换）技术通过VAN（增值网络）来实现企业间的商务交易。电子商务主要分为三大类：B2B（企业对企业的电子商务）、B2C（企业对消费者的电子商务）、C2C（消费者对消费者的电子商务）。

B2B（企业对企业的电子商务）：是指企业之间相互进行交易，诸如制造商与分销商、批发商与零售商之间所进行的交易。交易价格通常依据订单中的商品数量协商确定。

B2C（企业对消费者的电子商务）：是指企业基于购物软件中的产品列表与消费者之间进行的交易。

C2C（消费者对消费者的电子商务）：很多网站都免费提供产品分类广告、拍卖信息及论坛等，个体消费者可以通过在这些平台上诸如支付宝这样的网上支付系统轻松进行网上交易。淘宝网就是一个很好的例子，每天都会产生个人消费者之间的交易。

其他类型的电子商务，如G2G（政府对政府）、G2E（政府对员工）、G2B（政府对企业）、B2G（企业对政府）、G2C（政府对公众）、C2G（公众对政府），这些都涉及和政府之间的交易，如采购、报税、企业注册、更新许可证等。此外还有一些其他类型的电子商务，在此不再赘述。

虽然电子商务以 B2B 和 B2C 交易为主，但是人们越来越意识到 B2B 更具发展前景。全球 B2C 电子商务交易额从 1998 年的 78 亿美元增长到 2017 年的 2.2 万亿美元，与此同时，B2B 交易额则从 430 亿美元增加到了 20.3 万亿美元。毫无疑问，B2B 电子商务成为了全世界关注的焦点。

泛读材料

项目管理

项目管理是将知识、技能、工具和技术应用到项目活动中，从而满足项目的要求。项目管理是通过项目管理过程中的启动、计划、执行、监控和结束这些步骤的应用和集成来实现的。项目经理对项目目标的完成负有责任。

管理一个项目包括以下工作：

1）明确需求。

2）确立清晰的、可实现的目标。

3）平衡项目质量、规模、时间和成本这些竞争需求。

4）使项目的技术规范、计划和方法能够满足各利益相关者的不同要求和期盼。

需要注意的是，在整个项目生命周期中，由于事项逐步完善情况的存在和必然性，项目管理中很多过程是相互影响的。也就是说，当一个项目管理团队对一个项目了解得越多，这个团队就能更细致地管理此项目。

"项目管理"这一术语有时用来描述对项目管理，以及一些可重新定义为项目的、正在执行的操作管理的组织或管理方法，也可称之为"项目化管理"。采取这种方法的组织会把它的活动规划定义为符合项目定义的项目。最近几年，有一种趋势是使用项目管理在更多的应用领域中管理更多的项目活动。越来越多的组织正在使用"项目化管理"。这并不是说所有的操作都能或者都应该被组织成项目。"项目化管理"的采用还与采用接近项目管理文化的组织文化有关。尽管对项目管理的理解对于使用"项目化管理"的组织而言至关重要，对这一方法本身的细节性讨论不在此标准范围之内。

Appendix D 五大工程领域科技文献数据库列表

1. 机械工程领域

数据库名称	文献类型	提供机构/访问方式
Ei Compendex Web	期刊+会议+报告/文摘	The Engineering Information Inc（美国工程信息公司）
JCR Web	期刊/统计数据	ISI（美国科学信息研究所）
MIT's OpenCourse Ware	课程资料	http://ocw.mit.edu/OcwWeb/web/home/home/index.htm（免费）
National Academies Press（NAP）电子图书	图书/全文	http://www.nap/edu（免费）
OCLC ArticleFirst	期刊/题录	OCLC（联机计算机图书馆中心）
OCLC Ebooks	图书/题录+馆藏	OCLC
OCLC Electronic Collections Online（ECO）	期刊/题录+部分全文	OCLC
OCLC FirstSearch	多种文献/文摘	OCLC
OCLC PapersFirst	会议/题录	OCLC
OCLC Proceedings	会议/题录	OCLC
OCLC WorldCat（OCLC Online Union Catalog）	图书+网路+其他/题录+馆藏	OCLC
Clase Periodica	多种文献/题录	OCLC
GPO Monthly Catalog	政府出版物/题录	OCLC
ProQuest Digital Dissertations	学位论文/文摘+前24页（部分）	ProQuest Information and Learning
ProQuest 学位论文全文数据库	学位论文/全文	ProQuest Information and Learning
超星数字图书馆	图书/全文	http://pds.sslibrary.com（机构用户入口）
书生电子图书	图书/全文	http://edu.21dmedia.com
阿帕比（Apabi）电子图书	图书/全文	http://ebook.lib.apabi.com
万方数字化期刊全文数据库	图书/全文	http://g.wanfangdata.com.cn
中国期刊网全文数据库	图书/全文	http://www.cnki.net
维普中文科技期刊数据库	图书/全文	http://edu.cqvip.com
中国学位论文全文数据库	学位论文/全文	http://g.wanfangdata.com.cn

2. 电子工程领域

数据库名称	文献类型	提供机构/访问方式
EBSCO（ASP）	期刊/部分全文	EBSCO
Ei Compendex Web	期刊+会议+报告/文摘	The Engineering Information Inc

196

（续）

数据库名称	文 献 类 型	提供机构/访问方式
IEEE/IEE Electronic Library	期刊+会议+标准/全文	IEEE/IEE（美国电气电子工程师学会/英国电气工程师学会）
AIP、APS 期刊和 AIP 会议录	期刊/全文	American Institute of Physics（美国物理联合会）
The ACM Digital Library	多种文献/全文	http：//portal.acm.org（国外主站点） http：//acm.lib.tsinghua.edu.cn（清华大学图书馆镜像站点）
JCR Web	期刊/统计数据	ISI
MIT's OpenCourseWare	课程资料	http：//ocw.mit.edu/OcwWeb/web/home/home/index.htm（免费）
National Academies Press（NAP）电子图书	图书/全文	http：//www.nap/edu（免费）
Networked computer science technical reports library（NCSTRL）	科技报告/全文	http：//www.ncstrl.org（免费）
OCLC ArticleFirst	期刊/题录	OCLC
OCLC Ebooks	图书/题录+馆藏	OCLC
OCLC Electronic Collections Online（ECO）	期刊/题录+部分全文	OCLC
OCLC FirstSearch	多种文献/文摘	OCLC
OCLC PapersFirst	会议/题录	OCLC
OCLC Proceedings	会议/题录	OCLC
OCLC WorldCat（OCLC Online Union Catalog）	图书+网路+其他/题录+馆藏	OCLC
GPO Monthly Catalog	政府出版物/题录	OCLC
Clase Periodica	多种文献/题录	OCLC
ProQuest Digital Dissertations	学位论文/文摘+前24页（部分）	ProQuest Information and Learning
INSPEC	多种文献/文摘	The Engineering Information Inc
ProQuest Science Journals	期刊/部分全文	ProQuest Information and Learning
ProQuest 学位论文全文数据库	学位论文/全文	ProQuest Information and Learning
超星数字图书馆	图书/全文	http：//pds.sslibrary.com
书生电子图书	图书/全文	http：//edu.21dmedia.com
阿帕比（Apabi）电子图书	图书/全文	http：//ebook.lib.apabi.com
万方数字化期刊全文数据库	图书/全文	http：//g.wanfangdata.com.cn
中国期刊网全文数据库	图书/全文	http：//www.cnki.net
维普中文科技期刊数据库	图书/全文	http：//edu.cqvip.com
中国学位论文全文数据库	学位论文/全文	http：//g.wanfangdata.com.cn

3. 工业工程（含管理工程）领域

数据库名称	文献类型	提供机构/访问方式
Ei Compendex Web	期刊+会议+报告/文摘	The Engineering Information Inc
IEEE/IEE Electronic Library	期刊+会议+标准/全文	IEEE/IEE
ScienceDirect（Elsevier）	期刊/全文	Elsevier Science
ASTM	期刊/全文	ASTM（美国试验与材料协会）
AIP、APS 期刊和 AIP 会议录	期刊/全文	American Institute of Physics
John Wiley 电子期刊	期刊/全文	美国 John Wiley 公司
ISI Proceedings	期刊+会议/文摘	ISI
Kluwer Online	期刊/全文	Kluwer Academic Publisher
Springer Link	期刊/全文	Springer-Verlag
MIT's OpenCourse Ware	课程资料	http://ocw.mit.edu/OcwWeb/web/home/home/index.htm（免费）
NationalAcademiesPress（NAP）电子图书	图书/全文	http://www.nap/edu（免费）
High Wire Press	期刊/部分全文	http://www.highwire.org（免费）
OCLC Ebooks	图书/题录+馆藏	OCLC
OCLC Electronic Collections Online（ECO）	期刊/题录+部分全文	OCLC
OCLC FirstSearch	多种文献/文摘	OCLC
OCLC PapersFirst	会议/题录	OCLC
OCLC Proceedings	会议/题录	OCLC
OCLC WorldCat（OCLC Online Union Catalog）	图书+网路+其他/题录+馆藏	OCLC
Clase Periodica	多种文献/题录	OCLC
GPO Monthly Catalog	政府出版物/题录	OCLC
Wilson Select Plus	期刊/全文	OCLC
Transactions A	期刊/全文	英国皇家学会
ProQuest Digital Dissertations	学位论文/文摘+前24页（部分）	ProQuest Information and Learning
BIOSIS Previews	期刊+会议+图书/文摘	BIOSIS&ISI
Nature	期刊/全文	Nature Publishing Group
Chemical Abstract（CAonCD）	多种文献/文摘	CSA（加拿大标准协会）
ProQuest Science Journals	期刊/部分全文	ProQuest Information and Learning
EBSCO（MasterFILE Premie）	期刊+图书/部分全文	EBSCO
EBSCO（Newspaper Source）	报纸/部分全文	EBSCO
EBSCO（Vocational & Career Collection）	期刊+图书/部分全文	EBSCO

(续)

数据库名称	文献类型	提供机构/访问方式
Proquest Newspapers	报纸/全文	ProQuest Information and Learning
ProQuest学位论文全文数据库	学位论文/全文	ProQuest Information and Learning
超星数字图书馆	图书/全文	http://pds.sslibrary.com
书生电子图书	图书/全文	http://edu.21dmedia.com
阿帕比（Apabi）电子图书	图书/全文	http://ebook.lib.apabi.com
万方数字化期刊全文数据库	图书/全文	http://g.wanfangdata.com.cn
中国期刊网全文数据库	图书/全文	http://www.cnki.net
维普中文科技期刊数据库	图书/全文	http://edu.cqvip.com
中国学位论文全文数据库	学位论文/全文	http://g.wanfangdata.com.cn

4. 化学工程领域

数据库名称	文献类型	提供机构/访问方式
Ei Compendex Web	期刊+会议+报告/文摘	The Engineering Information Inc
EBSCO（ASP）	期刊/部分全文	EBSCO
ScienceDirect（Elsevier）	期刊/全文	Elsevier Science
AIP、APS期刊和AIP会议录	期刊/全文	American Institute of Physics
John Wiley电子期刊	期刊/全文	美国John Wiley公司
JCR Web	期刊/统计数据	ISI
ISI Proceedings	期刊+会议/文摘	ISI
Kluwer Online	期刊/全文	Kluwer Academic Publisher
MIT's OpenCourse Ware	课程资料	http://ocw.mit.edu/OcwWeb/web/home/home/index.htm（免费）
National Academies Press（NAP）电子图书	图书/全文	http://www.nap/edu（免费）
OCLC ArticleFirst	期刊/题录	OCLC
OCLC FirstSearch	多种文献/文摘	OCLC
OCLC PapersFirst	会议/题录	OCLC
OCLC Proceedings	会议/题录	OCLC
OCLC WorldCat（OCLC Online Union Catalog）	图书+网路+其他/题录+馆藏	OCLC
Clase Periodica	多种文献/题录	OCLC
GPO Monthly Catalog	政府出版物/题录	OCLC
ProQuest Digital Dissertations	学位论文/文摘+前24页（部分）	ProQuest Information and Learning
Proceedings A	期刊/全文	英国皇家学会
Transactions A	期刊/全文	英国皇家学会

（续）

数据库名称	文献类型	提供机构/访问方式
British Ceramic Transactions	期刊/全文	英国 Maney 出版公司
British Corrosion Journal	期刊/全文	英国 Maney 出版公司
Interdisciplinary Science Reviews	期刊/全文	英国 Maney 出版公司
International Materials Review	期刊/全文	英国 Maney 出版公司
Ironmaking and Steelmaking	期刊/全文	英国 Maney 出版公司
Materials Science and Technology	期刊/全文	英国 Maney 出版公司
Plastics Rubber and Composites	期刊/全文	英国 Maney 出版公司
Powder Metallurgy	期刊/全文	英国 Maney 出版公司
Science and Technology of Welding and Joinging	期刊/全文	英国 Maney 出版公司
Surface Engineering	期刊/全文	英国 Maney 出版公司
ProQuest 学位论文全文数据库	学位论文/全文	ProQuest Information and Learning
超星数字图书馆	图书/全文	http：//pds.sslibrary.com
书生电子图书	图书/全文	http：//edu.21dmedia.com
阿帕比（Apabi）电子图书	图书/全文	http：//ebook.lib.apabi.com
万方数字化期刊全文数据库	图书/全文	http：//g.wanfangdata.com.cn
中国期刊网全文数据库	图书/全文	http：//www.cnki.net
维普中文科技期刊数据库	图书/全文	http：//edu.cqvip.com
中国学位论文全文数据库	学位论文/全文	http：//g.wanfangdata.com.cn

5. 土木工程领域

数据库名称	文献类型	提供机构/访问方式
ASCE	多种文献/全文	http：//ascelibrary.aip.org
CSA Engineering Research Database	多种文献/文摘	http：//www.csa.com
Ei Compendex Web	期刊+会议+报告/文摘	Engineering Information Inc
JCR Web	期刊/统计数据	ISI
MIT's OpenCourse Ware	课程资料	http：//ocw.mit.edu/OcwWeb/web/home/home/index.htm（免费）
NAP 电子图书	图书/全文	http：//www.nap/edu（免费）
Congressional research service reports	科技报告/全文	http：//www.ncseonline.org/NLE/CRS（免费）
OCLC ArticleFirst	期刊/题录	OCLC
OCLC Ebooks	图书/题录+馆藏	OCLC
OCLC Electronic Collections Online（ECO）	期刊/题录+部分全文	OCLC

（续）

数据库名称	文 献 类 型	提供机构/访问方式
OCLC FirstSearch	多种文献/文摘	OCLC
OCLC PapersFirst	会议/题录	OCLC
OCLC Proceedings	会议/题录	OCLC
OCLC WorldCat（OCLC Online Union Catalog）	图书+网路+其他/题录+馆藏	OCLC
Clase Periodica	多种文献/题录	OCLC
ProQuest Digital Dissertations	学位论文/文摘+前24页（部分）	ProQuest Information and Learning
Proceedings A	期刊/全文	英国皇家学会
Transactions A	期刊/全文	英国皇家学会
ProQuest 学位论文全义数据库	学位论文/全文	ProQuest Information and Learning
超星数字图书馆	图书/全文	http：//pds.sslibrary.com
书生电子图书	图书/全文	http：//edu.21dmedia.com
阿帕比（Apabi）电子图书	图书/全文	http：//ebook.lib.apabi.com
万方数字化期刊全文数据库	图书/全文	http：//g.wanfangdata.com.cn
中国期刊网全文数据库	图书/全文	http：//www.cnki.net
维普中文科技期刊数据库	图书/全文	http：//edu.cqvip.com
中国学位论文全文数据库	学位论文/全文	http：//g.wanfangdata.com.cn

References

[1] 吕栋腾, 孙永芳. 实用科技英语 [M]. 西安：西北大学出版社, 2013.

[2] 杨承先, 杨璐维, 张琦. 现代机电专业英语 [M]. 2 版. 北京：清华大学出版社, 2012.

[3] 马佐贤. 数控技术专业英语 [M]. 3 版. 北京：北京理工大学出版社, 2016.

[4] 卜养玲. 数控专业英语 [M]. 北京：北京理工大学出版社, 2010.

[5] 田文杰. 科技英语教程 [M]. 西安：西安交通大学出版社, 2008.

[6] 王颖. 电子与通信专业英语 [M]. 北京：人民邮电出版社, 2006.

[7] 祝晓东, 张强华, 王璟. 电气工程专业英语实用教程 [M]. 2 版. 北京：清华大学出版社, 2012.

[8] 冀汶莉. 电子商务专业英语 [M]. 西安：西安电子科技大学出版社, 2011.

[9] 施平. 机械工程专业英语教程 [M]. 2 版. 北京：电子工业出版社, 2008.

[10] 王磊, 涂杰. 机电工程专业英语 [M]. 北京：冶金工业出版社, 2009.

[11] 董建国. 机械专业英语 [M]. 3 版. 西安：西安电子科技大学出版社, 2018.

[12] 杜慰纯, 宋爽, 李娜, 等. 信息获取与利用 [M]. 2 版. 北京：清华大学出版社. 2016.

[13] 董素音, 蔡莉静. 机电信息检索与利用 [M]. 北京：海洋出版社. 2008.

[14] 张振华. 工程信息检索与论文写作 [M]. 北京：清华大学出版社, 2009.

[15] 施平. 机械工程专业英语教程 [M]. 5 版. 北京：电子工业出版社, 2019.

[16] 谢小苑. 科技英语翻译 [M]. 北京：国防工业出版社, 2015.

[17] 余兴波, 霍金明, 顾晓琳. 电气信息工程专业英语 [M]. 北京：北京大学出版社, 2013.

[18] 李瑞, 王东辉, 洪梅. 机电专业英语 [M]. 北京：北京理工大学出版社, 2013.

[19] 潘宇, 钟子楠. 环境科学与工程专业英语 [M]. 哈尔滨：哈尔滨工业大学出版社, 2015.

[20] 李国厚, 赵欣. 自动化专业英语 [M]. 2 版. 北京：北京大学出版社, 2015.

[21] 沈星. 航空科技英语 [M]. 北京：北京理工大学出版社, 2015.

[22] 赵桂钦. 电子与通信工程专业英语 [M]. 北京：清华大学出版社, 2012.

[23] 甘辉. 汽车专业英语实用教程 [M]. 北京：机械工业出版社, 2010.

[24] 卜艳萍, 周伟. 计算机专业英语 [M]. 2 版. 北京：清华大学出版社, 2018.

[25] 刘曙光, 曹军义. 测控技术与仪器专业英语教程 [M]. 3 版. 北京：电子工业出版社, 2013.